T0262247

Fuel Injection Systems Handbook

Fuel Injection Systems Handbook

Edited by **Nicole Maden**

CLANRYE
INTERNATIONAL

New Jersey

Published by Clanrye International,
55 Van Reypen Street,
Jersey City, NJ 07306, USA
www.clanryeinternational.com

Fuel Injection Systems Handbook
Edited by Nicole Maden

International Standard Book Number: 978-1-63240-241-7 (Hardback)

Printed in the United States of America.

Contents

Preface

In this book, a descriptive account on fuel injection systems has been provided. It elucidates the central process that determines the development of internal combustion engines and performances of automotive vehicles. The book compiles original researches which focus on contemporary topics relevant to enhancing the injection phenomena per se and injection systems as the key components of the engine.

Various studies have approached the subject by analyzing it with a single perspective, but the present book provides diverse methodologies and techniques to address this field. This book contains theories and applications needed for understanding the subject from different perspectives. The aim is to keep the readers informed about the progresses in the field; therefore, the contributions were carefully examined to compile novel researches by specialists from across the globe.

Indeed, the job of the editor is the most crucial and challenging in compiling all chapters into a single book. In the end, I would extend my sincere thanks to the chapter authors for their profound work. I am also thankful for the support provided by my family and colleagues during the compilation of this book.

Editor

Introductory Chapter

A Look at Development of Injection Systems for High-Speed Direct Injection (HSDI) Diesel Engines

Kazimierz Lejda

Rzeszów University of Technology

Poland

1. Introduction

Current development of automotive industry is conditioned by the minimization of negative effects in relation to the environment. It results from the restrictions regarding exhaust emission limits which are introduced by the consecutive standards but it also guarantees market success of a given vehicle. Research carried out regarding customer preferences in UE countries have confirmed that during the purchase of a car they make decisions based on vehicle's ecological properties and the safety of use in road conditions. Next follow such criteria as performance and durability, which has so far been dominant. The obtainment of desirable operational indices by diesel engine depends to a high degree from the injection systems applied in these engines.

The recent years show the gradual and conspicuous progress in the development of structures and methods of control of injection systems in high-speed diesel engines designed for heavy duty vehicles. The stimulation of such progress is forced by the legislative norms and the market, as it often happens to technical equipment. In case of high-speed diesel engines these are the multi-aspect law regulations concerning overall environment impacts: the emission levels, noise spectrum and recycling compliance. On the other hand, the vehicle users expect the further minimization of operating costs, expressed mainly by the reduction of fuel consumption, maintaining the vehicle dynamic properties at the same time. The above circumstances caused significant acceleration of research and development work related to high-speed diesel engines. The improved environmental and operating parameters are determined by the process of formation and combustion of the fuel-air mixture. The process of creating the proper macro- and microstructure of the mixture is significantly affected by the fuel injection system; therefore the progress in this field is most conspicuous.

2. Requirements towards high-speed diesel engine injection systems

The operation of the piston combustion engine consists in the transformation of the chemical energy contained in the fuel supplied to the combustion chamber into the mechanical work received as the torque on the crankshaft end. In order to perform the chemical transformation into the mechanical work in a high-speed diesel engine, one should: I

- supply air to the combustion chamber,
- provide appropriate air-compression, in order to obtain temperature exceeding the fuel vapor self-ignition temperature,
- supply appropriate amount of fuel to the combustion chamber, correspondingly to the engine load,
- prepare appropriate structure of the fuel-air mixture,
- induce self-ignition of the mixture,
- transform the combustion gas pressure obtained as a result of combustion into mechanical work.

The supply of fuel in an appropriate amount and its penetration and distribution inside the combustion chamber are included in the injection system task. It directly creates the quality of the mixture prepared and the rate of the combustion process, which in turn translates into operating and environmental parameters as well as the engine operation economy. Out of the numerous factors influencing the quality of the combustible mixture preparation and the proper combustion that are directly dependent on the injection systems, the following can be enumerated:

- pressure and speed of the fuel injected,
- beginning, duration and end of injection,
- injection process and rate characteristics,
- amount of a single fuel injection,
- location of the fuel injected spray in the combustion chamber.

The significance of the factors varies and is mainly related to the injection system (indirect of direct injection), type of combustion chamber and requirements related to a specific engine type. The optimization of the high-speed diesel engines requires the matching of the injection system parameters to the loads and engine speeds, inherently related to the operation of traction engines and varying in real time. The accomplishment of this task requires precise control over the injection process and the parameters of its rate. The basic requirements made towards the injection systems, that are to comply with the currently applicable standardization and homologation regulations concerning the purity of exhaust and noise of operation as well as the reduced fuel consumption, may include:

- the possibility of controlling the injection process depending on the load, engine speed and engine temperature,
- the possibility of adaptive change of fuel amount (injection rate) depending on the load, engine speed and engine temperature,
- providing the optimum speeds of opening and closing of the nozzle needle and the needle lift values for the dynamically changing operating conditions of the engine,
- possibility of creating appropriately high pressures of injection adapted individually to the current operating conditions of the engine,
- providing the precise repeatability of the fuel amount injected in the particular cylinders and in the subsequent operating cycles, according to the current load, engine speed and thermal condition of the engine,
- the possibility of application to various types of engines resulting from specific applications (passenger cars, trucks, railed vehicles, stationary equipment, etc...).

In order to fulfill the requirements of the modern high-speed diesel engines, the injection systems must be electronically controlled. As the example of such systems is a widely used common-rail system. It is pressure-accumulation system, where the fuel pumping process is functionally separated and does not affect an injection rate. Thus, the accumulation systems provide a flexible rate of injection at the request of a particular engine design and the amount of fuel injected. They allow a freely-set multiple injections towards lowering noise level of the engine and reduction of the exhaust emissions. The pressure obtained in the accumulation systems usually exceeds 160 MPa.

3. Diesel injection systems development forecasts

Everything indicates that the development of high-speed diesel engines in the nearest future shall still be determined by the gradual reduction of the exhaust emissions and reduction of noise, together with a simultaneous increase in fuel economy. These trends concerning the exhaust emissions reduction since the introduction of Euro 1 standard can be illustrated by the consequent Euro levels. The data analysis proves the dramatic changes in legislation limitations within relatively short time intervals. Obtaining a reasonable compromise between fuel economy and exhaust emissions, particularly in the scope of NOx and PM emission, is the most difficult challenge at present. Fuel consumption restriction is related to the reduction of CO_2 emission – the supposed greenhouse effect cause.

The further simultaneous reduction of fuel consumption and pollutant contents in the exhaust gases may only be achieved through the introduction of state-of-the-art supply systems combined with the change of the rate of the combustion process. The leading automotive manufacturers carry out intensive work on reduction of power losses and achievement of general efficiency of diesel engines approaching 50%. Reduction of friction and thermal losses is very important for decrease of fuel consumption and the level of exhaust emissions, nevertheless the optimum rate of the combustion process shall always remain decisive for obtaining the demanded results. Therefore, the present works concentrate mainly on the issues related to examine as many physicochemical phenomena conditioning the proper combustion as possible. The theoretical analysis and experimental research indicate that the nature of the rate of the combustion process should be conspicuously modified. Providing the desired heat release rate is principally conditioned by the rate of the injection process, controlled by the injection systems. The application of a multiple injection is necessary here. At present, a standard five-phase fuel injection is applied. Each phase performs a specific function before and after the main injection. The pilot injection affects the noise reduction due to the reduction of the combustion pressure acceleration dp/dφ. The pre-injection intensifies the combustion of PM particles in the filters, whereas the secondary injection increases of NOx conversion in the DENOx catalytic reactor. Particularly the pilot injection and its time interval from the beginning of the main injection is significant for the combustion process. Due to the required limits in pressure growth and heat release rate during the initial combustion phase, especially at partial loads, the noise NOx and smoke emission can also be reduced.

The multiple fuel injection strategy requires the application of fast, new generation injectors. At the present stage of development such injectors are piezoelectric driven. They are characterized with shorter time of response than the electromagnetic injectors. The short response time enable to perform the multiple injections with very low level of fuel amount dispersion and with more accurate timing. Another characteristic of a piezoelectric injector are its much smaller dimensions. The further development of piezoelectric injectors (labeled as IV generation injectors) provides the possibility of application of variable injection rate and fuel injection pressure around 250 MPa. The introduction of such injector is signalized by Bosch (so-called HADI system – Hydraulically Amplified Diesel Injector).

4. Conclusion

The automotive market, throughout the last few years, is subject to conspicuous changes in the scope of clients' preferences. The sales of diesel engine cars gradually grow. Compared to the engine with spark-ignition, the diesel engine has a higher efficiency, lower fuel consumption and longer operation life. A diesel engine is much more easily adaptable to supercharging, which enables obtaining significant power gain without changing the stroke, displacement volume or engine speed. In a diesel engine that works with higher air-to-fuel ratio, the combustion is more efficient with lower carbon and hydrocarbons content in the exhaust gases. From the end user's point of view, the most important effect is the operation cost economy as the diesel engines consume less fuel. To meet the present and future demands, the automotive manufacturers work over new designs and technologies in diesel engines and offer a wide range of car models fitted with such drive units. The development of high-speed diesel engines is explicitly guided by the direct injection strategy and growth of the fuel injection pressure, exceeding 200 MPa. Such tendency forced intensive research and development works over new generations of injection systems. They must provide such parameters of mixture and the rate of the combustion process that would be able to meet the future requirements related to pollutant emissions in the exhaust gases and reduction of CO_2. These are the basic criteria determining the trends in the development and improvement of high-speed diesel engines and their injection systems but, in other hand, they signalize the scale of issues that the engineers and manufacturers of such devices must face. The classic injection systems cannot provide the proper rate of the injection process. Fulfilling the future requirements related to exhaust emissions and noise reduction combined with the fuel economy increase requires the absolute application of electronic control systems. This, in turn, is conditioned by the introduction of extended injection rate algorithms that may only be described by complicated 3D functional surfaces. The algorithms must also take into account the additional control parameters and functions and everything requires the strict application of adjusted values feedbacks.

Undoubtedly, the present state-of-the-art accumulation fuel systems provide great features and control precision of the injection process, but this advantage alone would not be able to meet further demands for green engines; it seems that development of direct injection strategy is closed to the systematic limitations. It is necessary to combine various progression trends, both in design and technology domains. As anticipated, the combustion engine shall be the basic drive source in the next 30-40 years, so the issue of development directions of the injection systems remains still an open question.

Section 1

Fuel Properties as Factors Affecting Injection Process and Systems

Multivariate Modeling in Quality Control of Viscosity in Fuel: An Application in Oil Industry

Leandro Valim de Freitas[1,2], Fernando Augusto Silva Marins[2],
Ana Paula Barbosa Rodrigues de Freitas[3],
Messias Borges Silva[2,3] and Carla Cristina Almeida Loures[2,3]
[1]*Petróleo Brasileiro SA (PETROBRAS)*
[2]*São Paulo State University (UNESP)*
[3]*University of São Paulo (USP)*
Brazil

1. Introduction

This chapter aims to present an alternative to quality control of the viscosity of two important fuels in the international scenery - aviation kerosene and diesel oil – by statistical multivariate modeling (Pasadakis et al., 2006).

Viscosity is one of the most important properties of fuels; it influences the circulation and the fuel injection in the operation of injection engines. Engines efficiency in the combustion process depends on this property.

Out of specification values can decrease the fuel volatilization, thus implying, in an incomplete combustion (Pontes et al., 2010). This physicochemical property can vary significantly with the modification of the cast during the processing of crude in a refinery (Figure 1), maintaining the same conditions of production control, which compromises the quality standards. This leads to the need to determine the viscosity or provide it as often as possible in lieu of performing the traditional point analysis in the laboratory that can take long time.

According to Dave et al. (2003), the use of field instruments in conjunction with statistical multivariate techniques to determine, in real-time, properties of the products is one way to optimize the operations of oil refining.

Each refinery has at least one primary distillation tower, where the components of crude oil are separated into different sections using different boiling points, and different arrangements of unit conversion. In general, the refining margin increases with the complexity of the refinery. Decisions about how to operate and monitor a refinery and how to build the units, are factors that provide competitive advantages to oil companies.

The hydrotreatment is a catalytic process that removes large amounts of sulfur and nitrogen from the distillation fractions (Fernández et al., 1995). The fluidized catalytic cracking (FCC) is a process widely used in a petroleum refining industry. It consists in cracking large molecules into smaller ones by high temperatures. Thus, heavy oils are converted into products with higher added value (IEA, n.d.). In addition, some refineries have used the coking unit to maximize the refining margin in the conversion of waste from distillation towers.

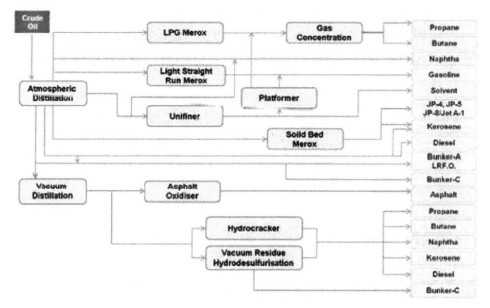

Fig. 1. Simplified process flow diagram of a refinery (Park et al., 2010)

Important products of refineries are the fuels for several kinds of existing engines. This way, fuel for jets is derived from petroleum and it is suitable for power generation by combustion in gas turbine engines for aircraft (see Fig.2). Jet fuel is produced by fractionation of petroleum by distillation at atmospheric pressure, with boiling range between 150 and 300 ° C, followed by finishes and treatments that aim primarily eliminate the undesirable effects of sulfur compounds, nitrogen and oxygen.

The viscosity of kerosene is limited to a maximum value to obtain a minimum loss of pressure in the flow at low temperature, as well as to allow using adequate spray nozzles for the fuel in order to improve the conditions of combustion. The viscosity property can significantly affect the lubricity of the fuel property, and, consequently, the life of the aircraft fuel pump.

The diesel fuel is a derivative of petroleum used in internal combustion engines compression to move motor vehicles (Fig. 3). It can also be used in marine engines and as a fuel for home heating. It is composed mainly of paraffinic hydrocarbons, and it is not desirable to the presence of olefins and aromatics. Its normal boiling range is 100 to 390 ° C, while the number of carbon atoms should be located between six and eighteen atoms.

The chemical composition of diesel oil directly affects its performance and is related to the type of oil used and with the adopted processes for their production in refineries.

Overall this product is composed of one or more cuts from the distillation of petroleum, and it can be added to other current refining processes, for example, the product obtained from catalytic cracking called light oil recycled.

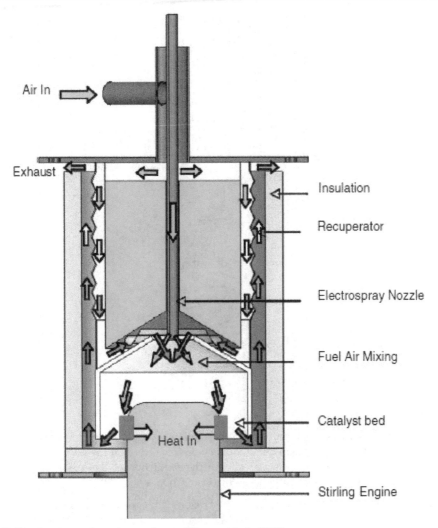

Fig. 2. Combustor and engine - schema (Gomez et al., 2007).

Kinematic viscosity of this product is an important property in terms of its effect on power systems and in fuel injection. Both high and low viscosities are undesirable since they can cause, among others, problems of fuel atomization. The formation of large and small droplets (low viscosity), can lead to a poor distribution of fuel and compromise the mixture air - fuel resulting in an incomplete combustion followed by loss of power and greater fuel consumption.

In Section 2, it will be presented a method for acquiring and construct the database. In Section 3, theoretical foundations of the statistical multivariate methods used. In Section 4, are presented the method application and the results for a real process of production. Finally, in Section 5, the conclusions reached are discussed.

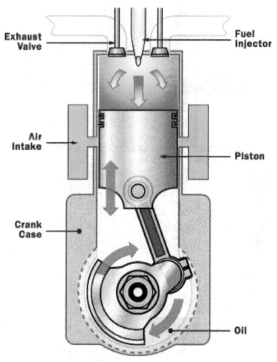

Fig. 3. Diesel engine (Marshall, 2002)

2. Acquisition data base

The multivariate analysis, using infrared signals, allow manipulate data absorption, called spectra, associated with more than one frequency or wavelength at the same time. In the oil industry, their applications are associated with the prediction of the quality of distillates such as naphtha, gasoline, diesel and kerosene (Kim, Cho, Park, 2000).

The chemical bonds of the type carbon-hydrogen (CH), oxygen-hydrogen (OH) and nitrogen-hydrogen (NH) are responsible for the absorption of infrared radiation, but are not very intense and overlap, creating broad spectral bands, that are correlated and difficult to interpret. However, a multivariate approach has proven quite adequate for modeling physical and chemical properties from samples of the input variables, it is known as absorbed infrared radiation (Behzadian et al., 2010).

The polychromatic radiation emitted by the source has a wavelength selected by a Michelson interferometer. The beam splitter has a refractive index such that approximately half of the radiation is directed to the fixed mirror and the other half is reflected, reaching the movable mirror and is therefore reflected by them. The optical path differences occur due to movement of the movable mirror that promotes wave interference. Nowadays, the instrumentation has introduced an improved Michelson interferometer, to develop the system "Wishbone", as illustrated in Figure 4. In this system, instead of having a movable mirror and a fixed mirror, both mirrors cubic move, tied to a single support.

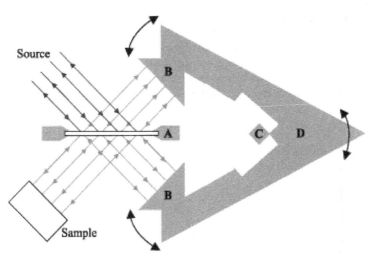

Fig. 4. "Wishbone" interferometric system employed in modern NIR spectrophotometers based on Fourier Transform. A, beam splitter; B, corner cubic mirrors; C, anchor, and D, "wishbone" (Pasquini, 2003)

An interferogram is obtained as a result of a graph of the signal intensity received by the detector versus the difference in optical path traveled by the beams. Then, like the Fourier Transform translates the recurring phenomenon in a series of sines and cosines (see Fig. 5), it is possible to transform the interferogram in a spectrum transmittance. The amount of radiation absorbed is determined by using the co-logarithm of the transmittance spectrum.

Fig. 5. Interferogram of radiation containing several wavelengths (adapted from Smith, 2011)

3. Multivariate analysis

The characterization of the mathematical models more adequate was performed by using the multivariate technique with partial least squares regression (PLS). This is an analysis technique where the original matrix of data is represented by factors or latent variables. Only the portion of the spectral data that correlates with the property assessed is included in this representation.

The first factor, calculated by a statistical program The Unscrambler®, has the highest correlation of spectral data with respect the property of interest. The residual spectrum, which is the original spectrum minus the proportion represented by the first factor, then the same statistical program evaluates it. Thus, the second factor has the highest correlation with the residual spectrum property. This procedure is replicated until each one of the important information, which has a correlation with the property under study, was represented by the factors or latent variables. Observe that some caution is needed to determine the appropriate number of factors, because an insufficient number of them will not include all the necessary spectral information and too many of them will add noise (see Fig. 6).

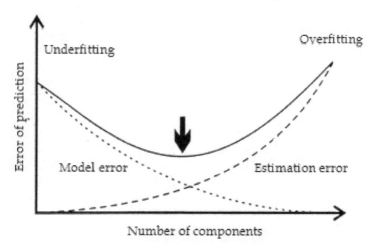

Fig. 6. Optimized number of components (Naes et al., 2002)

This regression model has the advantage of using the entire spectrum, be quick and offer a stable result. In addition, the regression using partial least squares, that use a number of factors less than the principal component regression (PCR) method, is more resistant to noise and in presence of weaker correlations.

The decomposition of the original variables on principal components can be represented by Equation (1), where for the k principal components, t is the value called score, that indicates the differences or similarities between samples, γ is the parameter that relates the original variables with the latent variable, it is called loading and represents how much a variable contributes to a major component and it considers the variation of the data:

$$tik = \upsilon k1Xi1 + \upsilon k2Xi2 + ... + \upsilon kqXkq \qquad (1)$$

So the principal components are related to the concentration, or property of interest, according to Equation (2), where h is the number of principal components used. The Fig. 7 represents the Equation (2) in matrix form:

$$Y = b0 + b1t\,i1 + b2t\,i2 + ... + bht\,ih + e1 \qquad (2)$$

The Fig. 8 illustrates the absorbance of three wavelengths. Observe that the first principal component PC1 is a linear combination of the absorbance values that representing the maximum variance between samples. The projection of the sample point on the axis of PC1 is the score of PC1 (see Fig. 9).

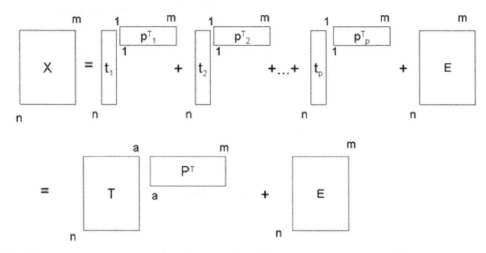

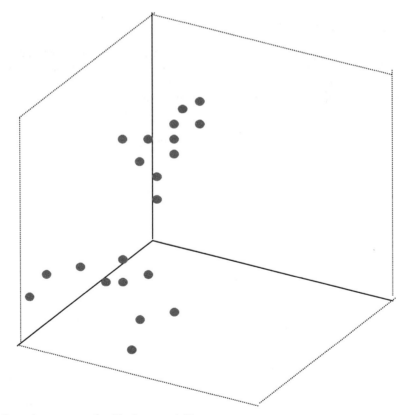

Fig. 7. Matrix representation of mathematical model

Fig. 8. Samples measured with three variables

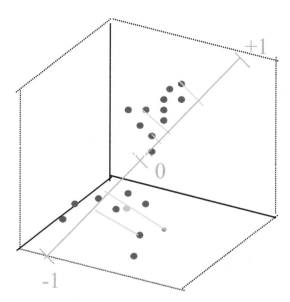

Fig. 9. First principal component in 3D

The PC2 must be orthogonal to PC1 and is positioned to capture the maximum residual variance. When all the data variability can not be explained by only one major component (red and green samples), a second PC is needed and so on. The score for the PC2 is obtained by projection, in a manner analogous to the previous situation (Fig. 10).

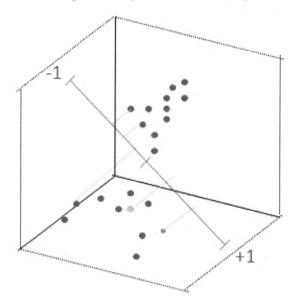

Fig. 10 . Second principal component in 3D

To obtain the remaining principal components, the procedure is the same. It be noted that for a set of 100 different wavelengths, for example, is not necessary 100 principal components to represent the data variability.

The variability of the spectrum can be compressed into a less number of principal components without significant loss of analytical information. After this compression, the scores are considered as independent variables in the regression to obtain the dependent variable y (concentration or physicochemical properties).

The PLS determines the principal components that are the best ones with respect to the variable y and that explain as best as possible the variable x (see Fig. 11).

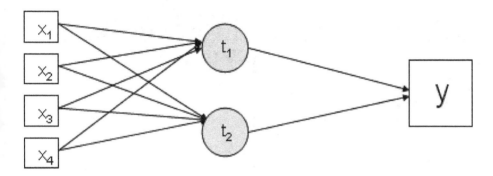

Fig. 11. Data compression in principal components

4. Proposed method, results and analysis

The first step is collect samples of kerosene and diesel oil in a reasonable period of time, to obtain a data set that best reflect all possible operational variability, as changes in the cast of oil and operating conditions of the units.

In the second stage experiments are performed to characterize, on a laboratory scale, the product, aiming to determining, from the samples, the real kinematic viscosity to be modeled. The samples were also subjected to infrared radiation.

In the third step the mathematical models were developed using The Unscrambler® and Excel® softwares, associating the information to absorbed infrared radiation with the physicochemical property. In the end, the model is implemented on an industrial scale for forecasting the viscosity in real time, providing to the production area, high power decision-making, and enabling increase the profitability of the blending process.

For each oil studied was developed a mathematical model with 800 input variables. To help determine the number of latent variables and minimize the residual variance was used the full cross-validation method, which is a mathematical algorithm able to gradually reduce the number of samples. In the sequence, a model constructed from the remaining samples is tested by comparing it with the true values of the samples excluded. The models are developed using The Unscrambler® program. Several forms of preprocessing were evaluated to obtain the minimum value of RMSECV (Root Mean Square Error of Cross Validation).

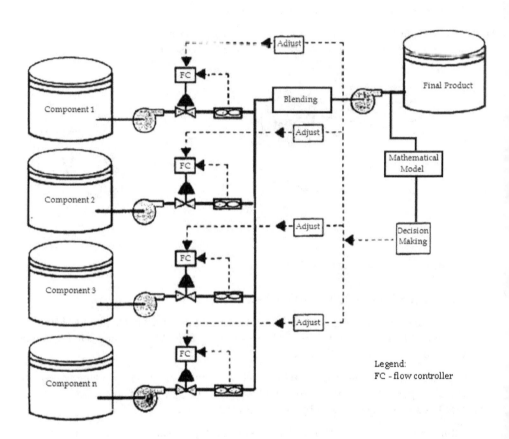

Fig. 12. Representation of online determination of viscosity by mathematical modeling (adapted from Early Jr, 1990)

The preprocessing that provided the best result was the first derivative of the second-degree polynomial proposed by Savitzky-Golay (Galvão et al., 2007), that highlight the differences between samples, contributing to the model can be used to explain the variance between them. The Fig. 13 and Fig. 14 show the original spectra for jet fuel and diesel, and the Fig. 15 and Fig. 16 show the same data after preprocessing by the above derivative.

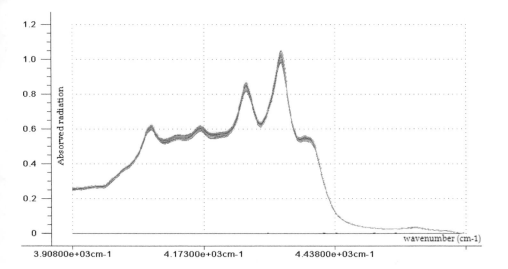

Fig. 13. Set of spectra of kerosene used in modeling

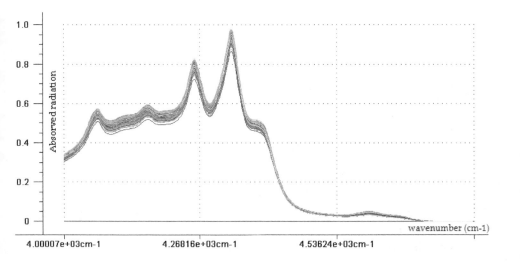

Fig. 14. Set of spectra of diesel used in modeling

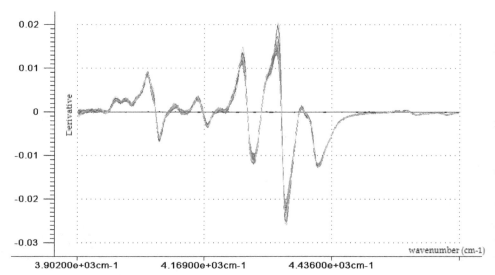

Fig. 15. Derivative (Savitzky-Golay) spectra of kerosene

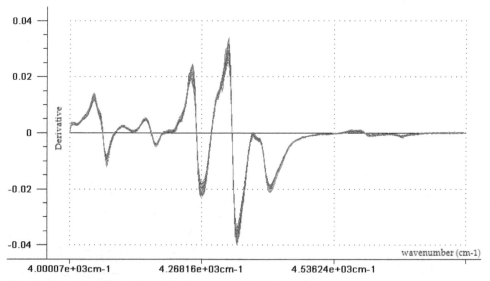

Fig. 16. Derivative (Savitzky-Golay) spectra of diesel

The method of the derivative of Savitzky-Golay smooths the spectrum by polynomial mobile. The derivative of the values of the absorbance as a function of wavenumber is calculated with the polynomial equation cited.

After the derivative of the spectrum, was determined the optimal number of latent variables for each product. In this way, for kerosene, were adopted four latent variables. Above this number, the explained variance decreases due to the incorporation of noise in the model, as shown in Table 1 and Fig. 17.

Latent Variable	Cumulative Variance Explained (%)
1	8.20
2	28.46
3	55.96
4	75.67
5	78.71
6	83.61
7	85.50
8	87.48
9	87.60
10	86.00
11	84.90
12	84.72

Table 1. Cumulative explained variance (%) versus latent variables (kerosene)

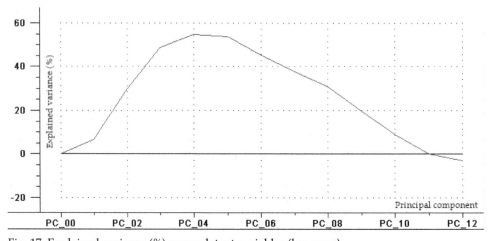

Fig. 17. Explained variance (%) versus latent variables (kerosene).

For the diesel, using the same procedure, were adopted eight latent variables, because above that number, there is no significant gain for explanation of the variance, as shown in Table 2 and Fig. 18.

Latent Variable	Cumulative Variance Explained (%)
1	8.20
2	28.46
3	55.96
4	75.67
5	78.71
6	83.61
7	85.50
8	87.48
9	87.60
10	86.00
11	84.90
12	84.72

Table 2. Cumulative explained variance (%) versus latent variables (diesel)

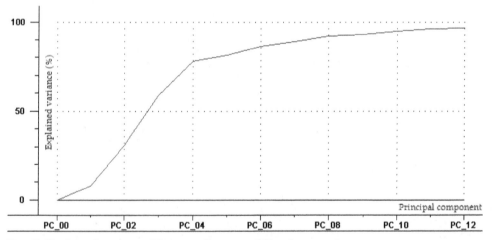

Fig. 18. Explained variance (%) versus latent variables (kerosene).

The outliers were eliminated observing the graphs of scores of the first two principal components and their influence (residual variance in Y versus leverage). The first two principal components captured the largest variability between the data and both for the samples of diesel and kerosene they are statistically close, as represented in the Fig. 19 and Fig.20.

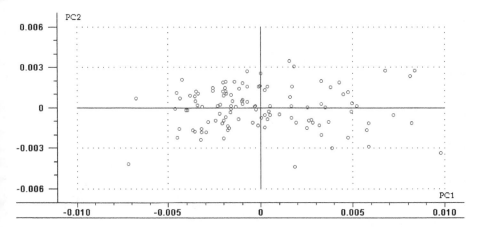

Fig. 19. Scores of the first two latent variables for kerosene

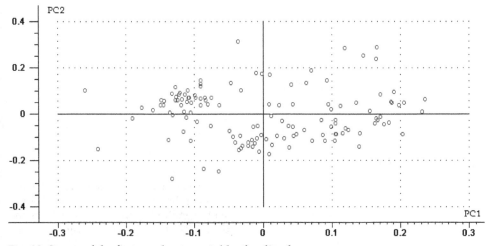

Fig. 20. Scores of the first two latent variables for diesel

Among the statistical tools used to detect outliers stands out the Student residues technique versus leverage. The leverage can be interpreted as the distance from the centroid of a sample data set. High values of leverage, means that the sample is located far from the mean and has a major influence in the model. The Student residues can be interpreted as the difference between the actual values and the values predicted by the model. For both products (jet fuel and diesel) there were no evident outliers, and it is also observed that the samples that have great influence have low residue (Fig. 16 and Fig. 17).

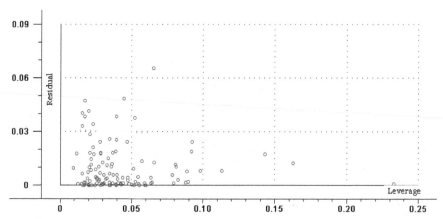

Fig. 21. Residual versus leverage for kerosene

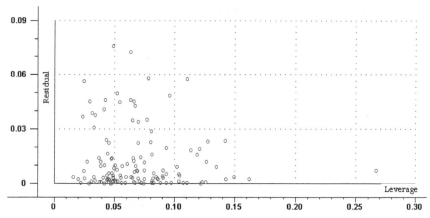

Fig. 22. Residual versus leverage for diesel

The correlation coefficient (r) indicates the degree of correlation between the estimated values and those obtained experimentally. In developing the model, the objectives were to minimize the RMSECV and maximize the coefficient of multiple determination (R^2) or the correlation coefficient (r). Graphs in Fig. 23 and Fig. 24 show the actual values versus predicted values along with the correlation coefficients.

The results of the models are statistically equivalent to those ones from laboratory methods. The Table 3 summarizes the results of modeling.

Property	Number of samples	Latent variables	RMSECV (cSt)	Correlation	Range (cSt)
Kerosene	115	4	0.09	0.889854	3.1 to 4.6
Diesel	131	8	0.11	0.959387	3.1 to 5.3

Table 3. Summary of modeling results

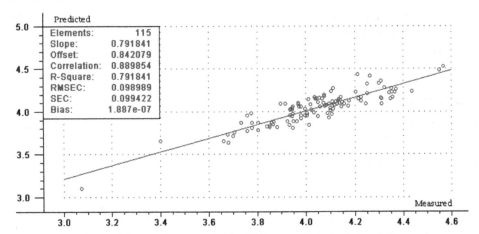

Fig. 23. Comparison of the results provided by PLS regression model and the results obtained by the reference laboratory (kerosene)

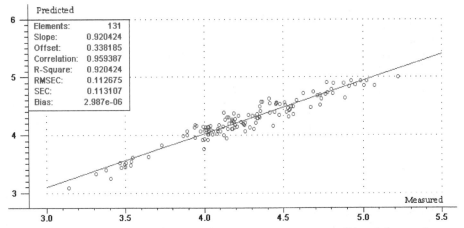

Fig. 24. Comparison of the results provided by PLS regression model and the results obtained by the reference laboratory (diesel)

5. Conclusions

It was possible modeling mathematically the kinematic viscosity of jet fuel and diesel oil by multivariate analysis. The values of the standard error of cross validation showed that the products meet the specifications.

It was possible to solve a real problem combining academia with industry. The results were used in an industrial plant at a refinery in Brazil and helped to speed up the decision-making in blending system, reduced process variability and increase the profitability of production.

There was a significant reduction of the analysis performed in the laboratory, because the proposed method is faster, more practical, does not generate chemical waste, minimize reprocessing, reducing costs and energy costs. In addition, better quality fuel reduces the impact of burning on the environment.

This chapter opens up other possibilities for assistance, such as simultaneous determinations of several other parameters of oil products.

6. References

Behzadian, M.; Kazemzadeh, R. B.; Albadvi, A. & Aghdasi, M. (2010). PROMETHEE : A comprehensive literature review on methodologies and applications. *European Journal of Operational Research*, Vol.200, (January 2009), pp.198-215, ISSN 0377-2217

Early Jr, P. L. (1990). In-line blending : using a PLC-based process controller. *ISA Transactions*, Vol.29, No.2, (November 1990), pp. 57-62, ISSN 0019-0578

Fernández, I.; Martinez, M. T.; Benito, A. & Miranda, J. L. (1995). Application of petroleum processing technology to the upgrading of coal syncrude. *Fuel*, Vol.74, No.1, (January 1995), pp. 32-36, ISSN 0016-2361

Galvão, R. K. H.; Araújo, M. C. U.; Benito, A.; Silva, E. C.; José, G. E.; Soares, S. F. & Paiva, H. M. (2007). Cross-Validation for the selection of spectral variables using the sucessive projections algorithm. *Journal of the Brazilian Chemical Society*, Vol.18, No.8, (December 2007), pp. 1580-1584, ISSN 0103-5053

Gomez, A.; Berry, J. J.; Roychoudhury, S.; Coriton, B. & Huth, J. (2007). From jet fuel to electric power using a mesoscale, efficient stirking cycle. *Proceedings of the Combustion Institute*, Vol.31, (June 2006), pp. 3251-3259, ISSN 1540-7489

International Energy Agency (IEA). (n.d.). Refining Processes, In: *Reliance Industries Limited*, 2.10.2011, Available from:
    < http://www.ril.com/html/business/refining_processes.html>

Kim, K.; Cho, I. & Park, J. (2000). Use of real-time NIR (near infrared) spectroscopy for the on-line optimization of a crude distillation unit, Proceedings of 12th Annual Computer Security Incident Handling Conference, pp. 1-4, Illinois, Chicago, USA, June 25-30, 2000

Marshall, B. (2002.). How diesel two-stroke engines work, In: *Howstuffworks*, 01.09.2011, Available from <http://auto.howstuffworks.com/diesel-two-stroke1.htm>

Naes, T.; Isaksson, T.; Fearn, T. & Davies, T. (2002). Multicollinearity and the need for data compression, In: *A user-friendly guide to multivariate calibration and classification*, NIR Publications, (Ed.), 19-24, ISBN 09528666, Chichester, England

Park, S.; Lee, S.; Jeong, S. J.; Song, Ho-Jun & Park, Jin-Won. (2010). Assessment of CO_2 emissions and its reduction potencial in the Korean petroleum refining industry using energy-environment models. *Energy*, Vol.35, No.6, (June 2010), pp. 2419-2429, ISSN 0360-5442

Pasadakis, N.; Sourligas, S. & Foteinopoulos, Ch. (2006). Prediction of distillation profile and cold properties of diesel fuels using mid-IR spectroscopy and neural networks. *Fuel*, Vol.85, No.7-8, (September 2005), pp. 1131-1137, ISSN 0016-2361

Pasquini, C. (2003). Near infrared spectroscopy : fundamentals, practical aspects and analytical applications. *Journal of the Brazilian Chemical Society*, Vol.14, No.2, (March 2003), pp. 198-219, ISSN 0103-5053

Smith, B. C. (2011). How an FTIR works, In: *Fundamentals of Fourier Transform Infrared Spectroscopy*, CRC Press, (Ed.), 19-54, Taylor & Francis, ISBN 978-142-0069-29-7, Boca Raton, United States

2

Role of Emulsified Fuel in the Present IC Engines – Need of Alodine EC Ethanol Corrosion Resistant Coating for Fuel Injection System

M. P. Ashok
Department of Mechanical Engineering, FEAT,
Annamalai University, Annamalai Nagar, Tamil Nadu,
India

1. Introduction

The present environment gets polluted day by day, due to the increase in industries, automobiles, etc., Eventhough, government has given strict rules and regulations against the environment pollution, practically there are lot of problems to implement the same by the users and production units, which leads to the result of global warming, acid rain, damaging of ozone layer affecting the green house effect and human health, etc. The pathetic information is that, the harmful pollutant does not allow the children also from their birth itself.

Due to the depletion of fossil fuel and its dangerous harmful emissions, the entire universe has given much attention for identifying an alternate fuel for the current existing engine. Also the fuel to be identified must have the property of fuel to the current existing engine and much free from the harmful pollution, emitted by IC (Internal Combustion) engine. Based on that, many research works have been carried out.

Present days emulsified fuel is much familiar for the IC Engines. Particularly ethanol based emulsion process fuels are much familiar, due to the depletion of fossil fuels. Entire universe has given much attention & many research works are going on based on the alternate fuel to the IC engines.

Based on that, emulsification technique is one of the possible approaches to identify an alternate fuel. Particularly /presently, the bio fuel from ethanol addition to the fossil fuels, play a vital role to run the current existing IC Engine, with best performance and fewer harmful emissions.

Presently the maximum blending of 50% Ethanol to 50% Diesel (50E: 50D), research work has been carried out by M.P.Ashok, academician, researcher and scientist {(in the field of Water-in-Oil) W/O–Emulsion} at Annamalai University, India, during the year 2007.

Based on the test result of Phase-I, it is possible to run the engine at 50% blending of ethanol, in the normal fuel injection system for a single cylinder, DI (Direct Injection) oriented engine.

Another point to be noted in this case is that, the engine gives better performance but emits more harmful emissions, at the higher rate for the emulsified fuel than the diesel fuel.

Further investigation has been carried out by the same author and identified the solution for reducing harmful emission by adding oxygen enriched additive Hydrogen Peroxide (H_2O_2) to the same emulsified fuel for increasing the cetane number of the fuel, with the selected emulsified fuel ratio of 50D: 50E.

Based on the test result of Phase-II, it is possible to run the engine at 50% blending of ethanol, in the normal fuel injection system for a single cylinder, DI (Direct Injection) oriented engine. Also, point to be noted in this case is that, the engine gives better performance and poor harmful emissions for the oxygen enriched additive added emulsified fuel than diesel and surfactant only added emulsified fuel.

Further investigation has been carried out by the same author, for selecting the best oxygen enriched additive among the three additives, namely Hydrogen Peroxide (H_2O_2), Di-Ethyl Ether (DEE) and Di-Methyl Ether (DME).

Based on the test result of Phase-III, it is possible to run the engine at 50% blending of ethanol, in the normal fuel injection system. Points to be noted in this case is that, the engine gives better performance and poor harmful emissions, for the oxygen enriched additive added emulsified fuel than the diesel and surfactant only added emulsified fuel.

Next research work has been carried out by the same author, introducing water as a fuel to the selected ratio of the emulsified fuel. Engine test report shows that, better performance and fewer emissions have been obtained.

Continuation of Phase-IV results, it is understood that, the engine could be run with the selected best ratio of the emulsified fuel along with water addition (small volume of water directly added to the emulsified fuel). The outcome of the test result have been given that, best in performance, lower emission obtained by the water, surfactant, oxygen enriched additive added emulsified fuel. Normally ethanol addition based fossil fuel cost is less than the normal fossil fuel cost. Further addition of water to the emulsified fuel, reduces the cost of the fuel much more. So, considering the cost economics, water added emulsified fuel is good for present engine operation.

The above mentioned outcome results, has given the details of best performance, poor harmful emissions and water addition using the best selected emulsified fuel ratio, based on ethanol addition. Also, it has been proved that the emulsified fuel is suitable for current existing engine.

But considering the part of corrosion, fossil fuel is already having the property of corrosion. Moreover, as ethanol is basically corrosive in nature, addition of ethanol fuel makes the engine components further more corroded. Also, some of the research works have already proved that ethanol fuel makes the engine parts to get more damaged.

Based on this, the present research work has been dealt for making Alodine corrosion coating for all the inner parts of the fuel injection system (normal fuel injection system used in the current (Internal Combustion) IC engine). The Alodine EC (Electro Ceramic) Coating the only and easy solution, which gives remedy for the corrosion caused by the emulsified fuel. This micro level Alodine coating based fuel injection system is much cheaper in cost

Role of Emulsified Fuel in the Present IC Engines – Need of Alodine EC Ethanol Corrosion Resistant
Coating for Fuel Injection System

27

and will increase the life span of the current fuel injection system (single and common system) available in the exiting engines, which is running under emulsified fuel category.

2. Role of emulsified fuel

2.1 Need of alcohol to the IC engine

Eugene Ecklund E., et.al, (1984), have given a detailed report on the concept of using alcohol fuels as alternative fuels to diesel fuel in diesel engine. They have also explained the different techniques for adding alcohol to the fossil fuels. In this case, blending of alcohol in the emulsion method has been clearly given and the merits and demerits of using alcohol as an alternate fuel are also explained. Their research has opened a whole new range of possibilities for using alcohols in transportation vehicles and has stated that the importance of this work will increase as the proportion of diesel-powered highway vehicles increases and as diesel fuel supply becomes more limited or degrades in quality.

2.2 Use of ethanol in CI engine

Ajav E. A. and Akingbehin O.A. (2002), have made a study on some of the fuel properties of ethanol blended with diesel fuel (six blends: 5, 10.15 . . .). Some properties have been experimentally determined to establish their suitability for use in CI engines. The results showed that both the relative density and viscosity of the blends decreased as the ethanol content in the blends has increased. Based on the findings of their report, blends with 5 &10% ethanol content are found to have acceptable fuel properties for use as supplementary fuels in diesel engines.

Alan.C.Hansen, et.al, (2004), have given a detailed review on ethanol-diesel fuel blends. They have stated that ethanol is an attractive alternative fuel because it is a renewable bio-based resource and it is oxygenated, thereby providing the potential to reduce the emissions in CI engines. Also, the properties and specifications of ethanol blended with diesel fuel have been discussed. Special emphasis has been placed on the factor of commercial use of the blends. The effect of the fuel on engine performance, durability and emissions has also been considered.

2.3 Use of surfactant to prepare the emulsified fuel

Santhanalakshmi.J and Maya.S.I, (1997), state about Span-80 and Tween-80 as the two non ionic surfactants, which could be used for preparing the emulsified fuel. Micellisation of surfactants in solvents of low dielectric constant differs from those in aqueous media due to the differences in the solute-solute, solute-solvent and solvent-solvent interactions. They have also explained about the solvent effects of the non ionic surfactants. They have stated that surfactants in non aqueous and non polar solution form reverse in order the micelles with hydrophilic core.

2.4 Selection of best emulsified fuel ratio

M.P.Ashok., et al., (2007) have conducted a research to identify the best ratio from the emulsified fuel ratios and compare with diesel fuel, based on its performance and emission characteristics. Water–in–Oil type emulsion method has been implemented to produce the emulsified fuel. Emulsified fuels have been prepared with different ratios of 50D: 50E (50 Diesel: 50 Ethanol - 100% Proof), 60D: 40E, 70D: 30E, 80D: 20E and 90D: 10E. From the investigation, it is observed that the emulsified fuel ratios have given the best result than diesel fuel. Also, 50D: 50E has given the best performance result than the other emulsified fuel ratios and diesel fuel. It has been observed that there is reduction in Smoke Density (SD), Particulate Matter (PM) and Exhaust Gas Temperature (EGT) with an increase of Oxides of Nitrogen (NOx) and Brake Thermal Efficiency.

2.5 Role of selected oxygen enriched additive in the emulsified fuel

Cherng – Yuan Lin and Kuo-Hua Wang, (2004) have given their report on the effects of an oxygenated additive in the emulsion field. Emulsified fuel characteristics have also been discussed, after adding the additives. They have shown the oxygenated additive to the diesel fuel which improves the combustion characteristics of the diesel engines. For the purposes of comparison, the emulsification characteristics of the two phases of emulsified fuels with additive have been analyzed. The chemical structures, HLB values and specific gravity of the surfactants Span-80 and Tween-80, have been given in detail. Also, they have stated that the efficiency and combustion properties of the CI engine have been improved by adding oxygenated additives in the emulsified fuel.

Mark.P.B.Musculus and Jef Dietz, (2005) have stated the effects of additives in the emulsion field on in-cylinder soot formation in a heavy duty DI diesel engine. Report states that the additives could potentially reduce in–cylinder soot formation by altering combustion chemistry. Chemical and physical mechanisms of the additives may affect soot formation in diesel engines. From the investigation, they have concluded that the effect of ignition delay on the soot formation and ethanol containing fuels display a potential for reduction of in – cylinder soot emission.

2.6 Role of best selected additive in the emulsified fuel (performance and emission)

M.P.Ashok., et., al., (2007) have studied about identifying an alternate fuel, for the current existing engines without any modification, with better performance and less emission. Emulsion technique is the best method to solve the above mentioned problems. Emulsified fuel with a surfactant is familiar nowadays. Addition of oxygen enriched additives in the emulsified fuels gives good results than the previous one. Usually NOx emission is high for the emulsified fuels, when compared with diesel fuel. But additive added emulsified fuels emit less NOx than diesel. Based on this, the present work has been carried out using oxygen enriched additives: Diethyl Ether (DEE), Hydrogen Peroxide (H_2O_2) and Dimethyl Ether (DME). The W/O type emulsion method is used to prepare the emulsified fuel. The test results have shown that DME has given the best performance and less emission, with the selected emulsified fuel ratio of 50D: 50E, comparing with the other two additives.

2.7 Role of emulsified fuel in the CI engine (performance and emission)

M.P.Ashok., et. al., (2007) have stated in their research work that various emulsified fuel ratios of 50D: 50E (50% Diesel No: 2: 50% Ethanol 100% proof), 60D: 40E and 70D: 30E have been prepared. Performance and emission tests are carried out for the emulsified fuel ratios and they have been compared with diesel fuel. The test results show that 50D: 50E has given

Role of Emulsified Fuel in the Present IC Engines – Need of Alodine EC Ethanol Corrosion Resistant
Coating for Fuel Injection System

29

the best result based on the performance and less emission than the other fuel ratios. By keeping the selected fuel 50D: 50E, the same performance and emission tests are conducted by varying their injection angles at 18°, 20°, 23° and 24°. The outcome shows better performance and less emission by the fuel 50D: 50E at 24° Injection Angle (IA). Further, ignition delay, maximum heat release and peak combustion pressure tests have been conducted. These results show that increase in IA decreases the delay period thus increasing the pressure obtained at the maximum output. Also, P-θ diagram is drawn between crank angle and cylinder pressure. The maximum value is attained by the fuel 50D: 50E at 24° IA. All the tests have been conducted by maintaining the engine speed at 1500 rev/min. The result shows that 50D: 50E ratio fuel has been identified as a good emulsified fuel and its better operation is obtained at 24° IA based on its best performance and less emissions.

2.8 Role of emulsified fuel in the CI engine (performance and emission)

M.P.Ashok., et. al., (2007) have studied about the best performance and less emission of 50% diesel and 50% ethanol [(50D/50E); 100% proof] emulsified fuel. Oxygen-enriched additive Dimethyl ether has been added to the selected best ratio of 50D: 50E emulsified fuel. Then, the performance and emission tests for diesel, 50D: 50E emulsified fuel ratio and oxygen-enriched additive-added emulsified fuel have been conducted. Finally, it has been found that the oxygen-enriched additive-added emulsified fuel has given the best performance and less emission when compared to the other two fuels. In comparison to diesel and the selected best ratio of the emulsified fuels, the oxygen-enriched additive-added emulsified fuel shows an increase in brake thermal efficiency and a decrease in SFC, PM, SD, and NOx.

Jae W.Park, et.al, [11] (2000) have done an experimental study on the combustion characteristics of emulsified fuel in a Rapid Compression and Expansion Machine (RCEM). Water–in–Oil emulsion type has been implemented and shows the best performance with respect to the better thermal efficiency. In the emission part, it is observed that NOx and Soot have been decreased. Also, the emulsified fuel has been characterized by a longer ignition delay and a lower rate of pressure rise in a premixed combustion.

2.9 Role of water added emulsified fuel in the CI engine (performance and emission)

Svend Henningsen, (1994), has investigated that NOx emission has been reduced by adding water to the emulsified fuel. The result shows that NOx is reduced with the addition of water, without deterioration in the SFC and the NOx behaviour is correlated, with the injection intensity as well as the water amount in the fuel. The report explains the result of the parameters such as injection valve opening, closing, duration, combustion starts and ignition delay. The concluding result is that the NOx emission and Specific Fuel Consumption (SFC) have been reduced considerably, because of water added to the emulsified fuel.

Wagner U., et.al, (2008), have described the possibilities of simultaneous in-cylinder reduction of NOx and soot emissions, for the DI diesel engines. They have stated that diesel engines with direct fuel injection give the highest thermal efficiency. Optimization of the injection process and the addition of water to the emulsified fuel are the two different possibilities for the reduction of NOx and soot emission, which have been discussed. Result of water addition gives increase in the value of thermal efficiency and reduction of NOx and soot emissions, when using the emulsified fuel. As the concept of water addition to the emulsified fuel leads to the reduction of peak combustion temperature, the NOx emission gets decreased. It concludes that the potential of water added emulsified fuel in the diesel combustion process has improved in thermal efficiency and reduction of especially NOx and soot emissions.

2.10 Role of corrosion water added emulsified fuel in the CI engine (performance and emission)

Teng Zhang and Dian Tang,, (2009), have discussed about the recent patents on corrosion resistant coatings. The materials of corrosion resistant components, e.g., metals and alloys, ceramics, polymers as well as composite materials, developed for environmental, economic and other concerns were discussed. In addition, the novel methods for forming the coatings, including the powder floating by vibration and the precursor gas, as well as some widely employed methods in the industrial applications were also included.

3. Procedure for the preparation of the emulsified fuel

3.1 With the help of surfactant

Normally Ethanol-in-Diesel emulsion fuel preparation method, diesel and ethanol are the dispersion and dispersed medium respectively. Hence, the dispersed medium is added slowly to the dispersion medium. The surfactant is used to reduce the interfacial tension between the diesel fuel and ethanol fuel. Here Tween-80 has been selected as surfactant, whose HLB (Hydrophile Lipophile Balance) value is 15. Based on the above, the selected surfactant reduced the interfacial tension between two fuels and producing the emulsified fuel. By varying different quantities of ethanol and diesel fuels at different ratios say 90D: 10E, 80D: 20E, 70D: 30E, 60D: 40E, 50D: 50E, with the variation of surfactant level, the emulsified fuel formed. But in and every cases the properties of the emulsified fuel have been changed.

In the Phase-I, for an example the best succeeded ratio of 50D: 50E has been prepared by adding 49.5% diesel fuel with ethanol fuel and the addition of surfactant Tween-80 of 1% by volume basis. Based on the above combination best emulsified fuel has been formed. All the addition of fuels, surfactants and other chemicals have been added by volumetric basis only.

3.2 With the help of surfactant and selected oxygen enriched additive addition

In phase-II, for the preparation of the emulsified fuel, initially the ethanol fuel 44% has been added with the diesel fuel 44% with the addition selected additive of Hydrogen Peroxide 11% has been added by volume basis. The above mentioned surfactant Tween-80 (1%) has been added to prepare the emulsified fuel.

3.3 With the help of surfactant and best selected oxygen enriched additive addition

Before preparing the emulsified fuel, the following oxygen enriched additives have been taken into account and finalized which additive is having higher rate of oxygen enriched properties for preparing the best emulsified fuel. For that, the following oxygen enriched additives of Hydrogen Peroxide (H_2O_2), Di-methyl Ether (DME) and Di-ethyl Ether (DEE) has been involved in the test. In phase-III, the outcome result shows that, the best selected oxygen enriched additive is DME, which is having higher rate of its oxygen enriched properties, when it is mixed with ethanol, diesel fuels and surfactant. Also the additive DME is having higher rate of oxygen enriched molecular condition, giving good stability of the emulsified fuel.

Based on that, ethanol, diesel, additive and surfactant have been added 45%, 45%, 9%and 1% respectively, added to prepare the emulsified fuel. Also the addition of oxygen enriched additive added emulsified fuel leads to give the best stability and increased life span of the of the emulsified fuel under the condition of Phase-I. Another main point to be considered is that, DME has the properties of highly volatile and easy evaporation based one. But considering the small addition by volume basis, it hasn't given any harmful one (For an Example: 1000ml (millilitre) total of emulsified fuel, the role of DME is 90 ml only, while preparing time. Considering the vaporization property of DME and liberation of oxygen atoms condition it doesn't give any harm).

3.4 With the help of selected surfactant, additive with water as fuel addition to the emulsified fuel

Already it is much familiar that, emulsified fuel cost is cheaper, when comparing with normal fossil fuel, because of the major addition of ethanol to the fossil fuel (Example: Diesel Fuel). So, in Phase-IV it has been considered that small quantity of water to be added to the emulsified fuel under the condition of Phase-III, for further reducing the cost of the emulsified fuel. Based on that 5% and 10% of water has been added to the emulsified fuel under the condition of Phase-III.

At one part the addition of water leads to reduce the performance of the engine and spoiling the emulsified fuel properties and considering the another part, water is also consists of oxygen atoms and having the evaporation property. So the water addition to the emulsified fuel doesn't majorly affect the property of the emulsified fuel (For an Example: out of 1000 ml, the water to be added only 5ml or 10ml). Base on that, the outcome result give the result of Phase-IV, with the condition if Phase-III.

3.5 Emulsified fuel producing machine operation with surfactant, additive and water fuel

Initially the required quantities of ethanol and diesel fuel are to be added as per the Water-in-Oil type emulsion type, along with the selected Surfactant. After adding all the above, the mixture is placed in a special type of mechanical stirrer, which has the specifications of 3-Phase, A.C. supply, 0-10000 rev/min variable speed, vertical motor having twin blades, helical shape attached with the vertical shaft of the motor, four numbers of zig- zag shaped blades which are fixed in the emulsified fuel containing drum vessel to get swirl motion for better mixing. After the required time interval, a good emulsion is formed due to the sharing effects produced by the helical blades of the shaft and fixed blades in the emulsified fuel vessel. Then selected best additive to be added along with the mixture of the emulsion, further the above mentioned action of the motor was started to get further best emulsified fuel. After getting the emulsified fuel, the required quantity of water fuel to be added to the emulsified fuel. The same Rapid Combustion and Expansion Machine (RCEM) action has been repeated, for the required time interval. At last the required emulsified fuel will be obtained. The stability period has been obtained 3 and 1/2 days for the prepared emulsified fuel. In all addition of the test, the required all chemicals have been added under volumetric basis condition.

3.6 Role of alodine EC ethanol corrosion resistant coating in the IC engine

Eventhough emulsified fuel is much cheaper in cost, by considering the cost economics further cheaper based on the water addition: ethanol, surfactant, additive, water are having basically corrosive in nature.

In normal day by day experience, any technical knowledge person could understand / come across that, the direct corrosion for the petro products used engine. In addition to the pert products, above mentioned emulsified fuel and its based chemicals are having the properties to damage the engine and engine parts easily. Based on the above statement, it is prove that, the life span of the engine gets reduced in faster rate.

So, emulsified fuel is having the advantage of being solution for the depletion of fossil fuels, could be used directly to the current existing engine, easily available, bio products based, it having its own disadvantages of corroding the engine parts very easily. Particularly the minute parts like fuel injector, fuel injection system, filter, piston, cylinder etc., gets corrode rapidly with faster rate / easily and the every where the problem will be raised in future, if the engine runs with emulsified fuel.

Based on this, in this present work, making the above mentioned engine parts to be saved from the chemical corrosion from petro products based emulsified fuel by giving a Alodine EC ethanol corrosion resistant coating, with minimum level thickness to the emulsified fuel holding / carrying engine parts and the parts which is kept in contact with emulsified fuel.

4. Experimental setup

The schematic diagram and the details of the test engine are given in the Figure 1 and Table 1 in details. Fuel flow rate is obtained by using the burette method and the airflow rate is obtained on the volumetric basis. NOx emission is obtained using an analyzer working on chemiluminescence principle.

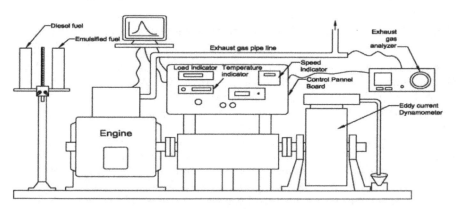

Fig. 1. Experimental Setup

The particulate matter from the exhaust is measured with the help of the micro high volume sampler. AVL smoke meter is used to measure the smoke capacity. AVL DIGAS 444 {DITEST} five-gas analyzer is used to measure the rest of the pollutants. A burette is used to measure the fuel consumption for a specified time interval. During this interval of time, how much fuel the engine consumes is measured, with the help of the stopwatch. Regarding the fuel injection system, MICO plunger pump type fuel injection system is used in this experiment. All the measurements are collected and recorded by a data acquisition system.

The specifications of the engine are given below:

Make	:	Kirlosker TV – I
Type	:	Vertical cylinder, DI diesel engine
Number of cylinder	:	1
Bore × Stroke (mm)	:	87.5 x110
Compression ratio	:	17.5:1
Cooling system & Fuel used	:	Water Cooling & Diesel
Speed (rpm)	:	1500
Rated brake power (kW)	:	5.4 @1500rpm
Fuel Injection pump	:	MICO inline, with mechanical governor, flange mounted
Injection Pressure (kgf/cm²)	:	220
Ignition timing	:	23° before TDC (rated)
Ignition system	:	Compression Ignition

Table 1. Specifications of the Diesel engine

The properties of Ethanol and Diesel No: 2 are given in table 2.

#	Ethanol (100% Proof)	Diesel No: 2
Chemical formula	CH_3CH_2OH	$C_{12}H_{26}$
Boiling point (°C)	78	180 to 330
Cetane number	8	50
Self ignition temperature (°C)	420	200 to 420
Stoichiometric air/fuel ratio (wt/wt)	9	14.6
Lower Heating value (kJ/kg)	27000	42800
Viscosity in centipoises at (20°C)	1.2	3.9
Specific gravity	0.783	0.894
Density (kg/m³)	794	830

Table 2. Properties of Ethanol and Diesel No: 2

5. Results and discussion

5.1 Phase – I selection of best emulsified fuel ratio (performance, emission and combustion)

Figure 2 shows the variation of brake thermal efficiency. All the emulsified fuel ratios have given the best efficiency than the diesel fuel. The difference in the value of the brake thermal efficiency at 5 kW between the emulsified fuel ratio of 50D: 50E and diesel fuel is 6.6%. This is due to more quantity of oxygen enriched air present in ethanol fuel than in diesel fuel. (Presence of volume of air in ethanol and diesel fuel is 4.3–19 and 1.5–8.2 respectively). The possible reason for this increase in efficiency is that, ethanol contains oxygen atoms, which are freely available for combustion, (Naveen Kumar., et.al., 2004). The oxygen present in ethanol generally improves the brake thermal efficiency, when it is mixed with neat diesel. Due to this reason, the brake thermal efficiency increases as concentration of ethanol is increased.

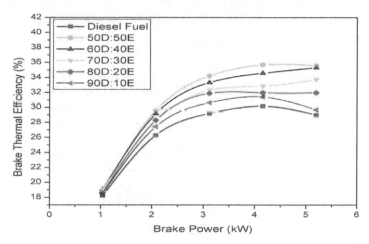

Fig. 2. Variation of Brake Thermal Efficiency

Figure 3. shows the variation of brake power verse SFC. SFC takes lower values for the emulsified fuels than the diesel fuel. This is because of the reduction of the energy content due to addition of ethanol, (Tsukahara, M and Yoshimoto, Y., 1992). Since, the energy content is low for ethanol, when it is mixed with diesel, it makes the emulsified fuel mixture to get poor in energy content. Also, the heating value of ethanol is lower, when compared to diesel. Due to this reason, the SFC is lower for the emulsified fuel ratio 50D: 50E.

As the brake thermal efficiency and SFC are inverse, the two basic parameters are most essential for a good performance of an engine. This could be achieved by the emulsified fuel ratio 50D: 50E. Therefore, the performance of the engine will be good, if it is run with emulsified fuel.

All the emulsified fuel ratios have taken less values of SD than the diesel fuel. The least value is taken by the emulsified fuel ratio 50D: 50E as shown in the figure 4. The reason is, addition of ethanol causes decrease in smoke level because of the better mixing of the air and fuel and increase in OH radical concentration, (K.A.Subramanian., A.Ramesh, 2001). Also, smoke emission of the ethanol–in–diesel fuel emulsion is lower than those obtained with neat diesel fuel because of the soot free combustion of ethanol under normal diesel engine operating conditions. Hence, as the ethanol concentration increases, the smoke density decreases.

All the emulsified fuels emit higher range of NOx than diesel fuel. Masahiro et al., (1997) have stated that generally alcohol/diesel fuel emulsion causes higher NOx emission because of the cetane–depressing properties of alcohol. Ethanol–diesel fuel emulsion causes high NOx emission because of low cetane number of ethanol. Low cetane number leads the fuel to increase the ignition delay and greater rates of pressure rise, resulting in higher peak cylinder pressures and high peak combustion temperatures. This high peak temperature increases NOx emission, (Masahiro Ishida,et.al., 1997). From the experiment, it is observed that as ethanol content increases, emission of NOx also increases.

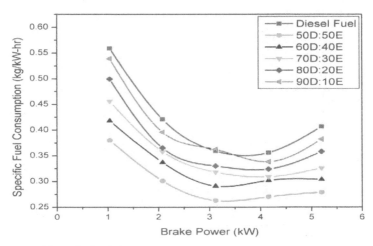

Fig. 3. Variation of Specific Fuel Consumption

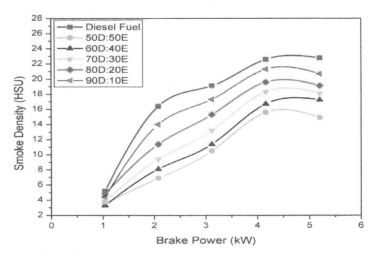

Fig. 4. Variation of Smoke Density

Also if ethanol mixes with any ratio with the diesel fuel, it emits more heat release. Considering the point of heating value, the difference between ethanol and diesel fuel is very small. This is adjusted by the higher latent heat of evaporation of ethanol. Even though the heating value of the ethanol fuel is less with the diesel fuel, the combustion takes place properly by the increased value of the latent heat of ethanol fuel. From this, it is understood that ethanol concentration is directly proportional to the heat release. At the rated output, heat release rate is the highest with ethanol–diesel operation due to enhancement of the premixed combustion phase. Normally, the rate of heat release depends largely on the turbulence intensity and also on the reaction rate, which is dependent on the mixture composition. Hence, 50D: 50E, 60D: 40E, 70D: 30E, 80D: 20E, 90D: 10E and finally the diesel fuel have taken the heat release rate based on the ethanol concentration.

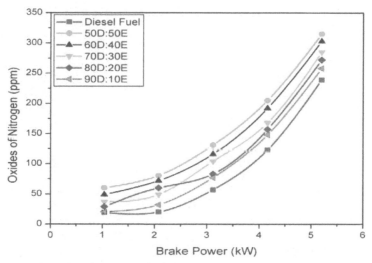

Fig. 5. Variation of Oxides of Nitrogen (NOx)

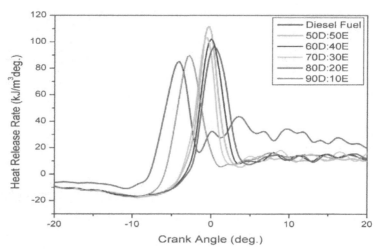

Fig. 6. Comparison of Heat Release Rate at 50% load

Figures 9 and 10 show the comparison of cylinder pressure at 50% and 100% load conditions. In general, there is no such significant change between the emulsified fuel and pure diesel. But there is a small rise in pressure caused by the emulsified fuel in both the cases. Basically, the pressure rise depends on the duration of the delay period. As the cetane number increases, the delay period decreases. Since ethanol has low cetane number, the ignition delay period is longer for emulsified fuel ratios (Cetane Number: for ethanol is 8 & diesel fuel is 50). This longer ignition delay helps to reach a high peak pressure to produce more work output during the expansion stroke. Due to this reason, the emulsified fuel ratios show higher pressure rise than diesel fuel. Also, the pressure rise is due to the amount of fuel involved in pre mixed combustion, which increases with longer ignition delay, (Tsukahara, M., et.al., 1982). Hence, the order is in the form of 50D: 50E, 60D: 40E, 70D: 30E, 80D: 30E, 90D: 20E and diesel fuel.

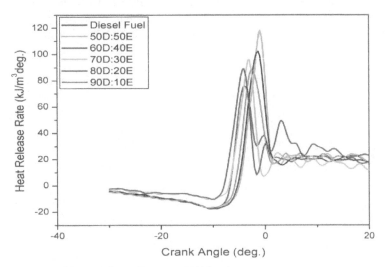

Fig. 7. Comparison of Heat Release Rate at 100% load

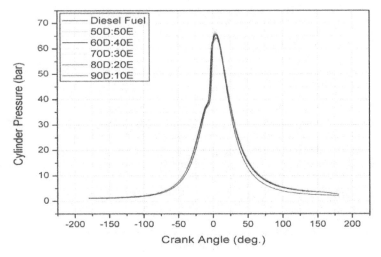

Fig. 8. P-θ for various fuels at 50 % Load condition

Figure 10 shows the comparison of brake thermal efficiency for all oxygen enriched additives added emulsified fuels DME, DEE and H_2O_2. Normally the oxygen enriched additives added emulsified fuels give greater brake thermal efficiency, because of their in higher cetane number. Higher cetane reduces the self-ignition temperature, which in turn reduces the delay period and results in smoother engine operation. Result of the longer ignition delay leads to a rapid increase in premixed heat release rate that affects brake thermal efficiency favorably. Also, the oxygen present in ethanol generally improves the brake thermal efficiency, when it is mixed with neat diesel (Dr.V.Ganesan). Based on this, the following oxygen enriched additives added emulsified fuels, take the role in the descending order of DME, DEE and H_2O_2. The maximum efficiency given by DME is 37.87%, at the maximum load condition. But considerable attention has to be given for the materials' compatibility and corrosiveness.

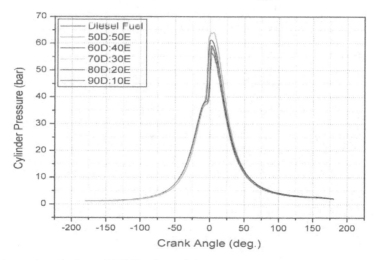

Fig. 9. P-θ for various fuels at 100 % Load condition

5.2 Phase – II & III best selected oxygen enriched additive and surfactant addition (performance and emission)

From figure 11, the SFC values are lower for all the fuels. Even though there is not much variation in the values, the order taken from minimum to maximum is the oxygen enriched emulsified fuels DME, DEE and H_2O_2 respectively. This is based on the energy content of the fuel. Normally, ethanol has less energy content than the diesel fuel. Based on this, the oxygen enriched emulsified fuel shows less value of SFC. Also, DME has the property of less energy content value than ethanol, (Cherng-Yuan Lin., et.al., 2004). Hence less SFC for the DME added emulsified fuel is found than in the other fuels. The least value obtained by DME, at the maximum load condition is 0.249 kg/kW-hr.

Figure 12 shows the comparison of SD, for all oxygen enriched additive added emulsified fuel. The order taken in the form of minimum to maximum is DME, DEE and H_2O_2 respectively. This is due to the better mixing of the air. Addition of ethanol causes decrease in smoke level and fuel and increase in OH radical concentration. The effect of fuel droplets vaporization plays a vital role with particular attention given for the oxygen content in the fuel as related to smoke density, (K.A.Subramanian. and A.Ramesh 2001). Because, oxygen enriched additives have more oxygen in nature, which lead to increase in OH radical concentration and oxygen content in the additive improves the fuel droplet size to get more vaporization. The least value obtained by DME is 46 HSU at maximum load condition.

Role of Emulsified Fuel in the Present IC Engines – Need of Alodine EC Ethanol Corrosion Resistant
Coating for Fuel Injection System

39

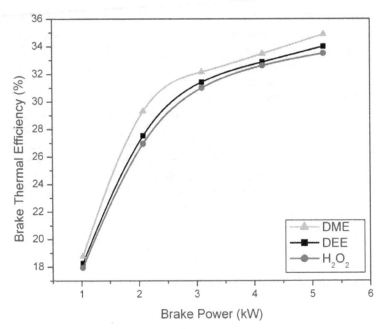

Fig. 10. Comparison of Brake Thermal Efficiency

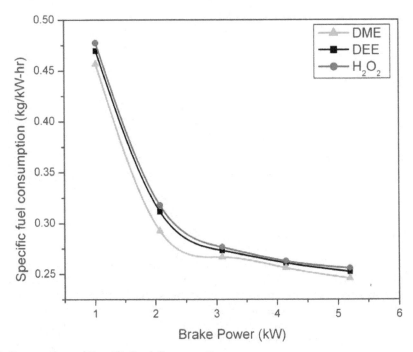

Fig. 11. Comparison of Specific Fuel Consumption

Figure 13 shows the comparison of the NOx emission, for various oxygen enriched additives added emulsified fuels, with the selected ratio of 50D: 50E. All the emulsified fuels emit higher range of NOx than diesel fuel. Masahiro et al., (1997) have stated that generally alcohol/diesel fuel emulsion causes higher NOx emission because of the cetane-depressing properties of alcohol. Normally, surfactant added emulsified fuels emit higher NOx than diesel fuel, because of its low cetane number, (M.P.Ashok 2007). Low cetane number leads the fuel to increase ignition delay and greater rates of pressure rise, resulting in higher peak cylinder pressures and high peak combustion temperatures. This high peak temperature increases NOx emission, (Masahiro Ishida 1997). But in the case of all oxygen enriched additives added emulsified fuels with the selected ratio of 50D: 50E less NOx is emitted It is because all the oxygen enriched additives have higher value of cetane number. Based on the higher cetane number the order takes place from minimum to maximum of DME, DEE and H_2O_2.

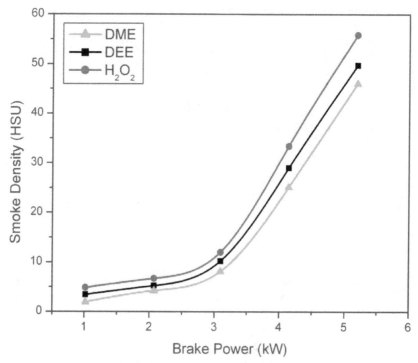

Fig. 12. Comparison of Smoke Density

Figure. 14 shows the comparison of maximum cylinder pressure, for different oxygen enriched additives DME, DEE and H_2O_2, with selected emulsified fuel ratio of 50D: 50E. Basically, the pressure rise depends on the duration of the delay period. As the cetane number increases, the delay period decreases. Since ethanol blending with oxygenated additives (quantity of additive added getting changed–by volume basis) has high cetane number, ignition delay period is shorter for additive added emulsified fuels, (Tsukahara, M. 1982). Based on this reason, the order taken from minimum to maximum is DME, DEE and H_2O_2.

Role of Emulsified Fuel in the Present IC Engines – Need of Alodine EC Ethanol Corrosion Resistant
Coating for Fuel Injection System

41

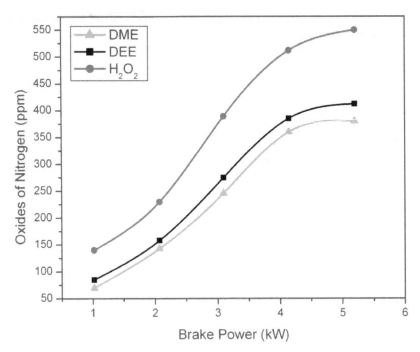

Fig. 13. Comparison of Oxides of Nitrogen

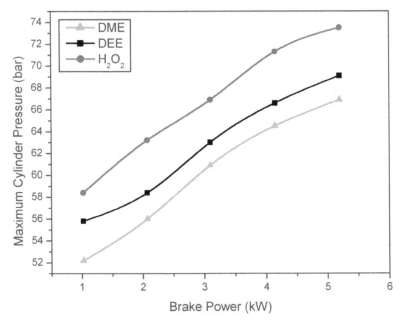

Fig. 14. Comparison of Maximum Cylinder Pressure

Comparison of heat release for the different oxygen enriched additives, with selected emulsified fuel ratio of 50D: 50E is shown in figure. 15. This is due to the higher and lower values of latent heat of evaporation of ethanol and diesel fuel respectively (Latent heat of evaporation: Ethanol–840 kJ/kg; Diesel–300 kJ/kg). At the rated output, heat release rate is the highest with ethanol-diesel operation due to enhancement of the premixed combustion phase, (Ajav. E.A, 1998). But the oxygen enriched additives DME, DEE and H_2O_2 added emulsified fuels have released minimum heat for the selected emulsified fuel ratio of 50D: 50E. In the case of oxygen enriched additives based emulsified fuels have more cetane number. Blending of additive and diesel leads to higher cetane number. Higher cetane number reduces the self-ignition temperature and hence emits less heat. Hence the oxygen enriched additive added emulsified fuels release less heat. Normally, the rate of heat release depends largely on the turbulence intensity and also on the reaction rate, which is dependent on the mixture composition. Based on these reasons, the order taken from maximum to minimum is H_2O_2, DEE and DME.

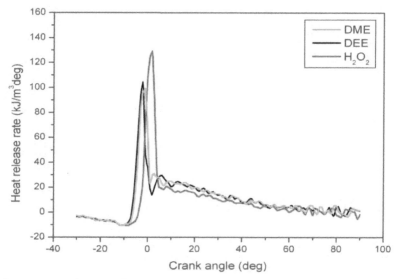

Fig. 15. Comparison of Heat Release Rate

5.3 Phase – IV addition of water fuel to the selected oxygen enriched additive and surfactant (performance and emission)

Figure 16 shows the variation of Brake Thermal Efficiency. There is no such output variation in the lower load condition. But at the middle and higher output level, there is a small variation. This is due to more quantity of oxygen enriched air present in ethanol fuel than in diesel fuel and the presence of oxygen content in water, (M.Abu-Zaid., 2004). Also, higher cetane number of diesel fuel leads to decrease in the delay period and causes reduced self-ignition temperature. Based on this, the variation is taken in the middle and the higher load conditions. The difference between the diesel fuel and the 10% water added emulsified fuel is 1.03% at 5.2 kW load condition.

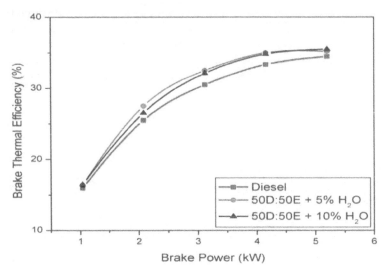

Fig. 16. Variation of Brake Thermal Efficiency

The variation of SFC is shown in figure 17. In this diesel fuel takes the maximum value than the remaining two fuels. This is because of the reduction of the energy content due to addition of water and ethanol, (Moses.C.A.,et.al., 1980). Already the ethanol fuel has less energy content and in addition to that if water is mixed with the emulsified fuel, it leads to very poor energy content of the fuel. Hence diesel fuel has attained the maximum value but the rest of places are attained by the emulsified fuels, according to the percentage of water addition. Also, the latent heat of vaporization of ethanol is high, when compared with diesel fuel. The addition of water reduces the latent heat of vaporization of the emulsified fuel.

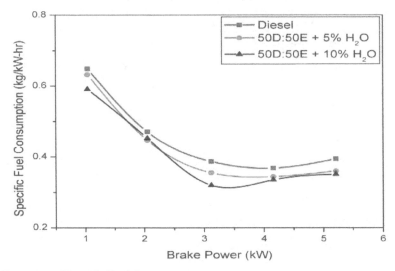

Fig. 17. Variation of Specific Fuel Consumption

Figure 18 shows the variation of SD. SD level increases for the emulsified fuel than diesel fuel due to poor mixing of air and fuel and increase in OH radical concentration, (Minoru Tsukahara., et.al., 1989). The same is higher for the emulsified fuel ratio 50D: 50E. The rest of the fuels are placed according to the order based on their OH radical concentration. The difference between the diesel fuel and the emulsified fuel (10% H_2O addition) emulsified fuel ratio is 14.4 HSU.

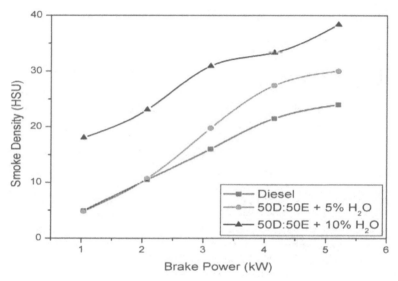

Fig. 18. Variation of Smoke Density

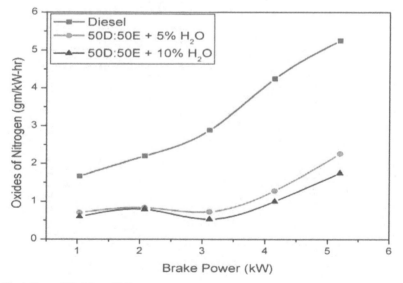

Fig. 19. Variation of Oxides of Nitrogen

Role of Emulsified Fuel in the Present IC Engines – Need of Alodine EC Ethanol Corrosion Resistant
Coating for Fuel Injection System

45

The variation of NOx is shown in figure 19. The low cetane depressing properties cause an increase in ignition delay and greater rates of pressure rise, resulting in high peak cylinder pressure and high peak combustion temperatures. The peak temperature always increases the NOx formation, (Masahiro Ishida and Zhi-Li Chen., 1994). Based on the above statement, the emulsified fuel emits more NOx. But in this experiment water is added to the emulsified fuel. Normally water addition reduces temperature. Hence NOx value gets decreased based on the peak combustion temperature reduction. From the above, the order is diesel fuel, emulsified fuel (5% H_2O addition) and finally emulsified fuel (10% H_2O addition). The difference between diesel fuel and the emulsified fuel (10% H_2O addition) at the maximum load condition is around 164 ppm. From the above, it is understood that NOx reduction is possible by using the water added emulsified fuel.

5.4 Phase–V Alodine EC ethanol corrosion resistant coating for fuel injection system and its parts dealing with emulsified fuel

The Alodine Electro Ceramic (EC) Corrosion Resistance Coating is the best solution for protecting the engine parts against corrosion. Presently the prepared emulsified fuel is more corrosive in nature and for that alodine EC corrosion resistant is the best solution. Application of alodine EC coating is cheaper and will be the best solution for corrosion. Particularly, the alodine EC coating could be applied to the minute parts of the engine. Manufacturers rely on EC to provide engine protection under a wide variety of extreme conditions, ranging from low temperature short trip service to extended high speed, high temperature operations.

Alodine EC also provides not only chemical protection but also wear resistance coating protection for intake manifolds, fuel injection system, fuel injection system pipe line, top of the piston, entire cylinder walls. It also reduces the friction in certain percentage; the performance of the engine gets increased.

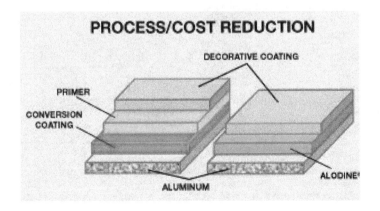

Fig. 20. Different layers of deposition based on Alodine EC coating

Fig. 21. Engine parts with alodine EC coating

Fig. 22. Engine parts after the Alodine EC coating

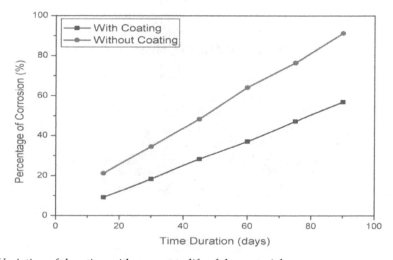

Fig. 23. Variation of duration with respect to life of the material

Role of Emulsified Fuel in the Present IC Engines – Need of Alodine EC Ethanol Corrosion Resistant
Coating for Fuel Injection System

47

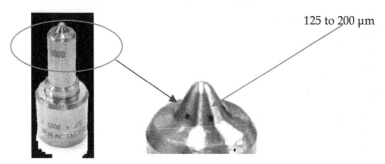

125 to 200 µm

Fig. 24. Alodine coated Fuel Injector and its Nozzle Part

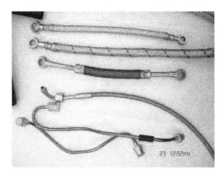

Fig. 25. Alodine coated Fuel Injection Fuel Lines

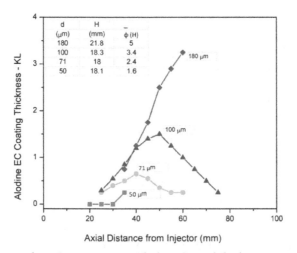

d (µm)	H (mm)	φ̄ (H)
180	21.8	5
100	18.3	3.4
71	18	2.4
50	18.1	1.6

Fig. 26. Axial Distance from Injector verses Alodine Coated thickness

From the above it is understood that Alodine EC Coating is much useful against corrosion. The thickness is around minimum and maximum of 25 and 250µm respectively. This would compromise the corrosion and increase the life span of the parts of the engine particularly fuel injection system.

Fig. 23 shows the variation of corrosion with standing capacity of the material verses duration days. In this case, alodine EC coated material and normal material have been kept in bath contact with Phase-IV condition of the emulsified fuel. Keeping the room temperature and moisture content, the above mentioned test has been carried out. After 90 days duration, both the materials have been undergone for the corrosion resistance test. Based on the test result, the alodine EC material gives maximum life then the ordinary material.

Also figure 24 & 25 show diagrams of Alodine EC coated fuel injector, edge of the nozzle part fuel injection hose lines. Figure 26 represents axial distance from the fuel injector, edge of the nozzle part verses Alodine EC Coating thickness given to the fuel Injector. It would indicate the different in thickness, diameter and height provided by the Alodine EC coated matrial. Also it shows that, the minute hole of the fuel injector gets Alodine coating with minimum thickness. Based on that entire fuel injection system gets safe against ethanol based emulsified fuel corrosion.

6. Future scope of the work

Further research work will be taken the study based on the performance, emission and combustion test along with alodine EC coated material of the engine parts, including fuel injection system.

7. Conclusions

- From Phase-I, it is well understood that Emulsified fuel 50E: 50D is the best emulsified fuel ratio for the CI Engine operation with increase in performance and NOx emission
- From Phase-II, it is understood that H_2O_2 additive added emulsified fuel under the ratio of 50D: 50E gives the best performance and less emissions of NOx.
- From Phase-III, it has been proved that DME is the best additive, when compared with the other two additives, which gives best performance and much poor emission.
- Phase-IV illustrates that water is also a fuel to the CI engine operation along with emulsified fuel.
- Finally Phase-V shows that Alodine EC coating is the best solution against corrosion, which has been caused by the emulsified fuel.

8. References

Abu-Zaid.M. (2004). *Performance of Single Cylinder, Direct Injection Diesel Engine using Water Fuel Emulsions,* Elsevier: Energy Conversion and Management 45 (2004) 697 – 705.

Ajav.E.A. & Akingbehin. O.A. (2002). *A Study of Some Fuel Properties of Local Ethanol Blended with Diesel Fuel* - CIGR Journal of Scientific Research and Development. EE 01 003. Vol. IV. March, 2002.

Ajav. E.A.; Singh. B. & Bhattacharya. T.K. (1998). *Performance of a Stationary Diesel Engine using vapourized Ethanol as Supplementary Fuel,* Bio Mass Bio energy 1998; 15(6): 493 – 502.

Alan.C. Hansen.; Qin Zhang. & Petter.W.L. Lyne. (2005). *Ethanol–Diesel Fuel Blends–A Review,* Elsevier – Journal of Bio Resource Technology – No: 96 (2005) Page No: 277 –285.

Role of Emulsified Fuel in the Present IC Engines – Need of Alodine EC Ethanol Corrosion Resistant
Coating for Fuel Injection System

49

Ashok.M.P. & Saravanan.C.G. (2007). *Selection of the Best Emulsified Fuel Ratio Compared to Diesel Fuel*, SAE Technical Paper No. 2007 – 01 – 2138.

Ashok.M.P. (2010). *Identification of Best Additive Using the Selected Ratio of Ethanol–Diesel Based Emulsified Fuel*, International Journal of Sustainable Energy, Taylor & Francis Publication, Paper No. gSOL-2010-0139.

Ashok.M.P. (2011). *Effect of Dimethyl Ether in a Selected Ethanol-Diesel Emulsified Fuel Ratio and Comparing the Performance and Emission of the same with Diesel Fuel*, Journal of Energy and Fuels, ACS Publications, Paper Number: ef-2011-007547.

Ashok.M.P. & Saravanan.C.G. (2007). *Combustion Characteristics of Compression Engine Driven by Emulsified Fuel under various Fuel Injection Angles*, Journal of Energy Resources and Technology, ASME Publication, Paper Number : JERT-06–1055.

Cherng-Yuan Lin. & Kuo–Hua Wang. (2004). *Effects of an Oxygenated Additive on the Emulsification Characteristics of two and three Phase Diesel Emulsions* – Elsevier – Journal of Fuel – Fuel 83 (2004) 507-515.

Eugene. E. Ecklund.; Richard Bechtold L.; Thomas J.; Timbario. & Petter W. McCallum. (1984). *State of the Art Report on the use of Alcohols in Diesel Engines*, SAE Technical Paper No: 840118.

Ganesan.V. (1998). *Internal Combustion Engines*, Publisher - Tata McGraw–Hill Book Company.

Jae W.Park.; Kang Y.Huh. & Kweon H.Park. (2000). *Experimental Study on the Combustion Characteristics of Emulsified Diesel in a RCEM*–Seoul 2000 FISITA World Automotive Congress – Paper No: F2000A073.

Mark.P.B.Musculus. & Jef Dietz. (2005). *Effects of Diesel Fuel Combustion – Modifier Additives on In–Cylinder Soot formation in a Heavy Duty DI Diesel Engine*, Sandia National Laboratories, Sandia Report, SAND – 2005 – 0189.

Masahiro Ishida.; Hironoku.Ueki. & Daisaku Sakauguo. (1997). *Prediction of NOx reduction rate due to port water Injection in a DI Diesel Engine*, SAE Paper No: 972961.

Masahiro Ishida. & Zhi-Li Chen. (1994). *An Analysis of the added water effect on NOx Formation in DI Diesel Engines*, SAE Technical Paper No: 941691.

Minoru Tsukahara.; Yasufumi Yoshimoo. & Tadashi Murayama. (1989). *W/O Emulsion Realizes Low Smoke and Efficient Operation of DI Engines without High Pressure Injection*, SAE Technical Paper No: 890449.

Moses . C.A.; Ryan.T.W. & Likos. W.E. (1980). *Experiments with Alcohol/ Diesel Fuel Blends in Compression Ignition Engines*. In: VI International Symposium on Alcohol Fuels Technology, Guaruja, Brazil.

Naveen Kumar.; P.B.Sharma.; L.M.Das. & S.K.Garg . (2004). *Ethanol – Diesel Emulsion as Diesel Fuel Extender*. SAE Technical Paper No: 2004 – 28 – 032.

Santhanalakshmi. J. & Maya.S.I. (1997). *Solvent Effects on Reverse Micellisation of Tween-80 and Span-80 in Pure and Mixed Organic Solvents* – Proceedings of the Indian Acad. Sci (Chem.Sci), Vol: 109, No: 1997: Page No: 27 -38.

Svend Henningsen. (1994). *Influence of the Fuel Injection Equipments on NOx Emission and Particulates on a Heavy – Duty Two – Stroke Diesel Engine Operating on Water–in–Fuel Emulsion*, SAE Technical Paper No: 941783.

Subramanian.K.A. & Ramesh.A. (2001). *Experimental Investigation on the Use of Water Diesel Emulsion with Oxygen- Enriched Air in a Di Diesel Engine*. SAE Technical Paper 2001-01-0205.

Teng Zhang. & Dian Tang. (2009). *Current Research Status of Corrosion Resistant Coatings,* 2009, Recent Patents on Corrosion Science, 2009, *1,* 1-5.

Tsukahara.M. & Yoshimoto, Y. (1992). *Reduction of NOx, Smoke, BSFC, and Maximum Combustion Pressure by Low compression ratios in a Diesel Engine Fueled by Emulsified fuel.* SAE Technical Paper No: 92046.

Tsukahara.M.; Murayama.T. & Yoshimoto. Y. (1982). *Influence of Fuel Properties on the Combustion in Diesel Engine Driven by the Emulsified Fuel,* Journal of the JSME, 1982, Vol. 25, No: 202, pp. 612 – 619.

Wagner U.; Eckert P. & Spicher U. (2008). *Possibilities of Simultaneous In–cylinder Reduction of Soot and NOx Emissions for Diesel Engines with Direct Injection.* International Journal of Rotating Machinery – Vol No: 2008: Article ID: 175956.

Section 2

Fuel Injection in ICE Versus Combustion Rate and Exhaust Emission

3

Simulation of Combustion Process in Direct Injection Diesel Engine Based on Fuel Injection Characteristics

Kazimierz Lejda and Paweł Woś
Rzeszów University of Technology
Poland

1. Introduction

Combustion engines are still the major propulsion devices for many mechanical equipment including mostly all automotive vehicles. Unfortunately, they negatively affects natural environment due to exhaust gas emission consisting of harmful compounds, like the carbon monoxide CO, unburned hydrocarbons HC, nitric oxides NO and NO_2 (altogether marked as NO_X), solid particles PM, and finally the carbon dioxide CO_2, that is to blame for the global warming phenomenon. All of them may cause human health deterioration or unwanted changes in the atmosphere in a large scale. As the examples, formation of the photochemical smog where hydrocarbons and oxides of nitrogen play the main role, or destruction of the ozone protective layer with participation of nitric oxide can be pointed here. There are also many other compounds in the exhaust gases which, besides their eco-toxicity, show also a serious carcinogenic action against people and animals, e.g. some hydrocarbons and particulate matter fractions.

The prevention-natured, legislative limitations of vehicle exhaust emission stemmed from these threats, together with the current and prospective growth of road transportation intensity, calls for continuous effort to develop vehicle powertrains that must be done both in design and technology domains. Hence the combustion engine improvement works have been spreading out within the space of last decades, and now they consider more and more factors. They pursue a simultaneous decreasing of harmful exhaust emission (CO, HC, NO_X, PM) and fuel consumption. Particularly, the thing is to cut down on CO_2 emission by increasing engine total efficiency. Fulfilling all above tasks encounters many problems. They contradict each other, what originates from complex physical and chemical interactions during the working cycle of piston engines, especially at combustion stage, where many phenomena combine together in the same time and area. For example, in direct injection engines simultaneously occurs: injection, fuel atomization and vaporization, induction of ignition or autoignition, fuel burning and many other chemical processes. All it takes only a few milliseconds. That is why improving exhaust emission parameters usually claims resignation from good fuel efficiency, and vice versa, fuel consumption decreasing escalates harmful emission. Hence, it is necessary to perform a lot experimental research in order to find an optimal solution. Unfortunately, they are generally complicated and expensive, but they could be successfully supported by numerical simulation. By the way, computational methods allow boundary-free analyzing and may narrow the range of further experiments. They are also much more time and cost-efficient than test bed investigations.

Fuel injection in combustion engines belongs to the most important working processes. Particularly, in direct injection (DI) engines, both gasoline and diesel one, it truly triggers and controls combustion, influencing all output engine performances together with exhaust emissions. Thus, improving of combustion engines should always consider fuel injection optimization, before other kinds of engine processes are being tested. Techniques harnessing mathematical simulations are the good ways to do this in the first stage. Such approach will be shown further.

2. The background of fuel injection process and its impact on combustion

Obtaining the desired operational engine parameters like break mean effective pressure (BMEP), overall engine efficiency, or environmental indexes (exhaust emissions, noise), essentially depends on the combustion rate, and previously on formation of flammable mixture. It particularly comes true in case of a direct injection engine. As far as a quality of combustible mixture is concerned, both the appropriate fuel atomization and fuel mixing with the air charge is important here, just as the precise metering of fuel amount injected in every cycle, and providing equal amounts of fuel for each cylinder of the engine. In direct injection engines, a fuel injection system has the main responsibility for creating proper fuel-air mixture, which can be characterized by an appropriate macro- and microstructure. Air swirl has a minor importance here, or even can make the whole process worsened due to cylinder wall wetting possibility.

On the whole, the mere combustion process is sensitive to many factors which may be classified into four groups (Wajand, 1988):

- physical and chemical properties of fuel, e.g. fractional and chemical composition, cetane/octane number, temperature of fuel (auto)ignition, etc;
- structural properties of the engine, e.g. combustion chamber shape, main engine dimensions, compression ratio, materials used, etc;
- operation and regulation conditions of the engine: rotational speed of the crankshaft, engine load, ignition (or start of injection) timing, etc;
- fuel injection system layout that generates fuel delivery process with the specific fuel spray parameters; it influences directly combustion rate, as is discussed above.

These considerations indicate that the fuel injection parameters are important factors affecting combustion process and consequently the engine parameters, what has been also verified by experimental studies (Kuszewski & Szlachta, 2002; Zabłocki, 1976).

A process of fuel injection into the engine cylinder might be described by parameters related to injection rate and spray characteristics. The set of injection rate parameters consists of:

- start of injection (SOI) timing referred to crankshaft position angle [deg],
- injection duration referred to crankshaft rotation angle [deg],
- mean flow rate of the fuel in a whole injection duration [mm^3/deg],
- actual fuel injection rate (instantaneous fuel flow velocity) [mm^3/deg],
- maximum fuel flow velocity [mm^3/deg].

Presently, the pressure-accumulative fuel systems with electronically controlled injectors are widely used, e.g. Diesel common rail (CR) one. In such systems, the injection rate essentially depends on two parameters: the shape of electrical signals in the injector and the hydraulic-mechanical characteristic of the injector. The change of pressure in the fuel storage is so small and affects fuel injection rate so little that it can be neglected in simulation works on engine operation.

The fuel spray characteristics may be described by the following parameters:

- spray tip velocity and penetration,
- spray tip angle,
- fuel atomization quality expressed by mean diameter of droplets and its dispersion,
- distribution of fuel mass along and across the spray,
- equivalent mean droplets size: linear, areal, volumetric, areal-volumetric (Sauter).

Preparation of flammable mixture in a direct injection engine becomes involved not only with strategy of fuel delivery, but also with areal distribution of fuel in the cylinder space. It should be noticed that a lot of factors influencing the fuel injection rate also plays an important role in fuel atomization quality, resulting in fuel mixing with air inside the cylinder.

3. Demonstration of combustion model

All mathematical models of combustion engine working cycle can be sorted as follows (Rychter & Teodorczyk, 1990):

1. considering dimensions, we have:
- zero-dimensional models,
- quasi-dimensional models,
- multi-dimensional models;
2. considering a number of recognized zones, we have:
- one-zone models,
- two-zone models,
- multi-zone models.

Above segmentation defines a model complexity and fidelity in representation of real processes in the model. It is also connected with complication in mathematical tools used for simulation. There are a lot of examples which combines the models according to the above segmentation (Khan et. al., 1973; Patterson, 1994, 1997; Rychter & Teodorczyk, 1990). The fundamental problem in choosing a proper type of the model is to find a compromise between accuracy and intellectual labor involved to describe all physical phenomena. A priority here is the goal of analysis. As a rule, for comparative and/or quantitative research, a simplified model can be used with receiving good results; for qualitative investigations more precise model should be worked out instead.

In a preliminary analysis toward model formulating, a number of physical and chemical processes that occur during the injection, combustion and exhaust pollutant formation were taken into consideration. The latest theoretical and experimental results were regarded. A great effort was made to include to the analysis all phenomena that have a major impact on the various processes modeled, so as their actual nature would be reproduced. Thanks to this, the model enables to simulate the effect of many structural and operational factors on the engine performance, including detailed emissions.

As a result, a two-zone quasi-dimensional model has been developed. Comparing to single-zone models, the own one is characterized by a much more accurate description of the phenomena in a combustion chamber. This attribute greatly emphasizes the scientific and utilitarian aspect of such a solution. In addition, the model permits to be extended easily of additional, computational blocks. In this respect, the proposed method of analysis exemplifies an important cognitive value and is rarely found in the literature.

Below, an application of worked out, two-zone, quasi-dimensional model of combustion for direct injection diesel engine will be presented. Splitting the combustion chamber into two zones for models of such type of the engines makes a fidelity in representation of phenomena proceeded inside the cylinder much precise, although it complicates mathematics.

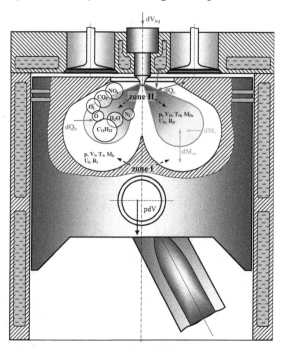

Fig. 1. The scheme of physical and chemical processes proceeded in a combustion chamber of direct injection (DI) diesel engine (Woś, 2008)[1]

As is shown in the Fig. 1, inside the cylinder of volume V and pressure p, at the end of intake stroke there is a fresh air charge of a mass M_{ch}, and at the moment determined by the start of injection angle, an initial fuel quantity begins to be injected. The fuel volume flow rate is $dV_{inj}/d\varphi$. Here, $d\varphi$ means the increment of an independent variable that is the crankshaft angle. A part of the fuel begins to evaporate with the rate equals to $dM_v/d\varphi$. It forms the spray cones of total volume V_{II}, consisted of fuel-air mixture. Through that, the

---

[1] The denotations are explained in the chapter body.

combustion chamber is divided onto two zones: the first one (I) - consisting of the rest of fresh charge, and the second one (II) - drawn by fuel-air mixture boundaries. After short time of autoignition delay τ_o, the process of evaporated fuel combustion gets set on and runs with the burning rate equals to $dM_h/d\varphi$. It generates a heat flux $dQ_h/d\varphi$ that is supplied into the zone II. Between both zones (I and II), a mass transfer process occurs ($dM_{ex}/d\varphi$), and between cylinder walls and both zones a heat transfer process occurs also with the rate $dQ_c/d\varphi$. The whole system gives an elementary mechanical work equals to $pdV/d\varphi$. Except of fundamental combustion reaction, the other free selected ones can be implemented (dissociations, pollutant formation) in order to check various engine output parameters.

3.1 The model core based on thermodynamic theory

In relation of above physical model, a mathematical model of engine working cycle was formulated with taking some indispensable assumptions into consideration. The essential equation for energy conversion inside the cylinder is differential equation of the first law of thermodynamics for open systems:

$$\frac{dU}{d\varphi} = \frac{\partial Q}{d\varphi} - p\frac{dV}{d\varphi} + \frac{dH}{d\varphi} \tag{1}$$

where:

U–internal energy of the system [J],

Q–heat delivered to/derived from the system [J],

V–system volume [m³],

p–system pressure [Pa],

H–enthalpy delivered to/derived from the system [J],

φ–crank angle [deg].

Above equation is valid for both zones of the elaborated model, but it must be developed further in order to calculate temperature change in both zones. According to the assumptions taken in the physical model, we can write as follows (detailed evaluation can be found in (Woś, 2008)):

for heat fluxes:

$$\frac{\partial Q_I}{d\varphi} = -\frac{\partial Q_{cI}}{d\varphi}$$
$$\frac{\partial Q_{II}}{d\varphi} = -\frac{\partial Q_{cII}}{d\varphi} + \frac{\partial Q_h}{d\varphi} - \frac{\partial Q_v}{d\varphi} \tag{2}$$

for mass transfers:

$$\frac{dM_I}{d\varphi} = \frac{dM_{ex}}{d\varphi}$$
$$\frac{dM_{II}}{d\varphi} = \frac{dM_v}{d\varphi} - \frac{dM_{ex}}{d\varphi} \tag{3}$$

and for enthalpy fluxes:

$$\frac{dH_I}{d\varphi} = h_{ex} \cdot \frac{dM_{ex}}{d\varphi}$$

$$\frac{dH_{II}}{d\varphi} = h_v \cdot \frac{dM_v}{d\varphi} - h_{ex} \cdot \frac{dM_{ex}}{d\varphi} \quad (4)$$

where:

I, II-subscripts referred to zone I and II, in order,

Q_c-heat of cooling [J],

Q_h-heat generated by combustion [J],

Q_v-heat consumed by vaporizing fuel [J],

M_{ex}-mass transferred between both zones [kg],

M_v-mass of evaporated fuel [kg],

h_{ex}-specific enthalpy of transferred mass; it is specific enthalpy of I or II zone depending on direction of mass flow [J/kg],

h_v-specific enthalpy of fuel vapor [J/kg],

Total internal energy of any thermodynamic system can be expressed by multiplying specific internal energy u and system mass M. Thus, we can also differentiate this multiplication, what gives:

$$\frac{dU}{d\varphi} = \frac{d(M \cdot u)}{d\varphi} = u \cdot \frac{dM}{d\varphi} + M \cdot \frac{du}{d\varphi} \quad (5)$$

If we consider that specific internal energy u for various compounds mixture can be calculated: $u = \sum_i (u_i \cdot g_i)$; then, assuming $\sum_i \left(u_i \cdot \frac{dg_i}{d\varphi} \right)$ as near to null, we will get:

$$\frac{du}{d\varphi} = \frac{dT}{d\varphi} \cdot \sum_i \left(g_i \cdot \frac{\partial u_i}{\partial T} \right) \quad (6)$$

where:

u_i-specific internal energy for an "i" component [J/kg],

g_i-mass fraction of an "i" component in a whole system [kg/kg],

T-temperature of the system [K].

Substitution of all above equations into the fundamental equation (1) for both zones will give a system of two differential equations with three unknowns: $dT_I/d\varphi$, $dT_{II}/d\varphi$, and $dM_{ex}/d\varphi$:

$$\begin{cases} \left[\sum_i \left(u_{Ii}g_{Ii}\right) - h_{ex}\right]\dfrac{dM_{ex}}{d\varphi} + M_I \sum_i \left(g_{Ii}\dfrac{\partial u_{Ii}}{\partial T}\right)\dfrac{dT_I}{d\varphi} = \dfrac{\delta Q_I}{d\varphi} - p\dfrac{dV_I}{d\varphi} \\[4mm] \left[h_{ex} - \sum_i \left(u_{IIi}g_{IIi}\right)\right]\dfrac{dM_{ex}}{d\varphi} + M_{II} \sum_i \left(g_{IIi}\dfrac{\partial u_{IIi}}{\partial T}\right)\dfrac{dT_{II}}{d\varphi} = \dfrac{\delta Q_{II}}{d\varphi} - p\dfrac{dV_{II}}{d\varphi} - \left[\sum_i \left(u_{IIi}g_{IIi}\right) - h_v\right]\dfrac{dM_v}{d\varphi} \end{cases}$$

$$(7)$$

The other differentiates, such as $dQ_I/d\varphi$, $dQ_{II}/d\varphi$, $dV_I/d\varphi$, $dV_{II}/d\varphi$, $dM_v/d\varphi$, can be calculated with use of independent submodels. To resolve above system algebraically, $dM_{ex}/d\varphi$ must be eliminated and expressed by other known components. To do that, we can use an overall assumption that the pressure p in both zones is always equal in value:

$$p_I = p_{II} \tag{8}$$

According to the ideal gas law equation of Clapeyron, it also means that:

$$\frac{M_I \cdot R_I \cdot T_I}{V_I} = \frac{M_{II} \cdot R_{II} \cdot T_{II}}{V_{II}} \tag{9}$$

where the symbols refer to both zones, such as subscript indicates, and they mean as follows:

M–mass of the zone [kg],

R–universal gas constant for a whole zone [J/(kg·K)],

T–average temperature of the zone [K],

V–zone volume [m³].

Going ahead, at any time the mass of the first zone is the sum of initial mass of fresh air charge M_{ch} and mass transferred M_{ex}. Similarly, for the second zone it is a mass of evaporated fuel M_v from which the transferred mass M_{ex} must be subtracted. Then transferred mass can be evaluated as follows:

$$M_{ex} = \frac{M_v \cdot R_{II} \cdot T_{II} \cdot V_I - M_{ch} \cdot R_I \cdot T_I \cdot V_{II}}{R_I \cdot T_I \cdot V_{II} + R_{II} \cdot T_{II} \cdot V_I} \tag{10}$$

To differentiate it relatively to the crank angle variable φ, we receive a formula for component $dM_{ex}/d\varphi$ as a function expressed by the other differentiates:

$$\frac{dM_{ex}}{d\varphi} = f\left(\frac{dT_I}{d\varphi}, \frac{dT_{II}}{d\varphi}, \frac{dV_I}{d\varphi}, \frac{dV_{II}}{d\varphi}, \frac{dM_v}{d\varphi}\right) \tag{11}$$

Now, replacing the component $dM_{ex}/d\varphi$ in the system of equations (7) with the above function we receive a new system of two differential equations with only two unknowns: $dT_I/d\varphi$, $dT_{II}/d\varphi$, i.e.:

$$\begin{cases} A \cdot \dfrac{dT_I}{d\varphi} + B \cdot \dfrac{dT_{II}}{d\varphi} = C \\[4mm] D \cdot \dfrac{dT_I}{d\varphi} + E \cdot \dfrac{dT_{II}}{d\varphi} = F \end{cases} \tag{12}$$

where A, B, C, D, E, F contains expressions of known variables, which can be evaluated by use of independent submodels and/or separated formulas. In this shape of the system, the unknowns $dT_I/d\varphi$, $dT_{II}/d\varphi$ can not be calculated numerically yet. Such methods need explicit from of the equations. To get it, the system (12) has to be transformed (solved algebraically) relatively to $dT_I/d\varphi$, $dT_{II}/d\varphi$, which mean the variables now. For instance, applying the method of Cramer determinants we will get:

$$\begin{cases} \dfrac{dT_I}{d\varphi} = \dfrac{B \cdot F - E \cdot C}{B \cdot D - E \cdot A} \\[2mm] \dfrac{dT_{II}}{d\varphi} = \dfrac{C \cdot D - A \cdot F}{B \cdot D - E \cdot A} \end{cases} \tag{13}$$

Above computer simulation friendly form of equations can be already implemented into the numerical calculation package and allow program running. Obviously, this core model has to be added with necessary sub-models describing other phenomena like heat transfer, fuel injection, fuel atomization and evaporation, ignition delay, combustion rate, combustion products formation, etc. The chosen ones will be shown further.

3.2 Fuel injection model

Modeling of the fuel injection in the combustion chamber space is one of the most difficult issues in all simulation works regarding the processes in reciprocating combustion engines. This is caused mainly by limited expertise knowledge in this field. Thus, in simulation works covering the combustion chamber space, a representation level of fuel injection submodel, as well fuel evaporation and combustion one is assumed with consideration of total accuracy of the whole model. More complex, spatial mathematical description should be used only in the cases where the injection process is the main essence of modeling.

In the current study, a number of simplifying assumptions in description of the fuel injection process have been made. Nevertheless, they were tailored to the level of accuracy in whole zero-dimensional model layout. According to the preliminary analysis, it has been assumed that the distribution of fuel density in the sprays generated is the same in all directions; next, the shape of sprays is characterized by a constant tip angle, and spray microstructure is described by the mean droplet diameter according to Sauter definition (SMD – Sauter Mean Diameter) and it is uniform throughout the entire space of fuel jet.

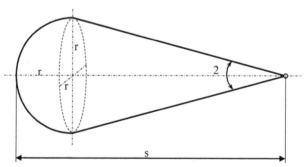

Fig. 2. A model for single fuel spray geometry (Woś, 2008)

According to the physical model layout that is shown in Fig. 1, the geometry of fuel injection sprays defines the volume of the zone of fuel-air mixture signed as zone II. A following simplified single spray cone geometry model has been adapted (Fig. 2).

Spray tip penetration s and tip angle 2α are the principal parameters. They allow calculating the volume of zone II (V_{II}) by multiplying the volume of elementary spray and number of sprays generated by the injector, i.e. the number of holes in the sprayer i:

$$
\begin{aligned}
V_{II}(\varphi) &= i \cdot \left[\frac{1}{3}\pi \cdot \left(\frac{s(\varphi)}{ctg\,\alpha + 1} \right)^2 \cdot s(\varphi) \cdot \left(1 - \frac{1}{ctg\,\alpha + 1} \right) + \frac{2}{3}\pi \cdot \left(\frac{s(\varphi)}{ctg\,\alpha + 1} \right)^3 \right] = \\
&= i \cdot \frac{1}{3}\pi \cdot s(\varphi)^3 \cdot \left[\frac{1}{\left(ctg\,\alpha + 1\right)^2} + \frac{1}{\left(ctg\,\alpha + 1\right)^3} \right]
\end{aligned}
\tag{14}
$$

Equation (14) expresses an instantaneous volume of the zone II that varies within the injection duration. According to the assumptions made, the change of zone II volume (V_{II}) just depends on a spray tip penetration increasing $ds/d\varphi$:

$$
\begin{aligned}
\frac{dV_{II}}{d\varphi} &= i \cdot \pi \cdot s(\varphi)^2 \cdot \left[\frac{1}{\left(ctg\,\alpha + 1\right)^2} + \frac{1}{\left(ctg\,\alpha + 1\right)^3} \right] \cdot \frac{ds}{d\varphi} = \\
&= i \cdot \pi \cdot s(\varphi)^2 \cdot \left[\frac{1}{\left(ctg\,\alpha + 1\right)^2} + \frac{1}{\left(ctg\,\alpha + 1\right)^3} \right] \cdot v_s(\varphi)
\end{aligned}
\tag{15}
$$

where the symbols used in above equations, both (14) and (15), mean:

$V_{II}(\varphi)$–total volume of combustion zone II at specified crankshaft angle position [m^3],

$dV_{II}/d\varphi$–change of total volume of zone II [m^3/deg],

i–number of holes in the sprayer [–],

$s(\varphi)$–spray tip penetration at specified crankshaft angle position [m],

$v_s(\varphi)$–spray tip velocity at specified crankshaft angle position [m/deg],

α–a half of spray tip angle [rad],

φ–independent variable: crankshaft angle position [deg].

Empirical formulas have been used for further analysis. They are based on criterion numbers that are widely used in fluid mechanics. And so, the relationship of spray tip penetration s and velocity v_s is given by the following formulae (Orzechowski & Prywer, 1991):

$$
s(\varphi) = \sqrt{\frac{d_r \cdot w_0}{\sqrt{2} \cdot a_u}} \cdot \frac{1}{6 \cdot n} \cdot \left| \varphi - \varphi_{inj} \right|
\tag{16}
$$

$$
v_s(\varphi) = \frac{ds}{d\varphi} = \frac{d_r \cdot w_0}{2 \cdot \sqrt{2} \cdot a_u \cdot s(\varphi)}
\tag{17}
$$

where:

$s(\varphi)$, $v_s(\varphi)$–spray tip penetration [m] and velocity [m/s] at crankshaft angle position φ,

d_r–spraying hole diameter [m],

n–crankshaft rotational speed [1/min],

φ–independent variable: the current crankshaft angle position [deg],

φ_{inj}–crankshaft angle position at start of injection [deg],

w_0–an initial velocity of fuel spray tip left the injector [m/s],

$$w_0 = \mu \cdot \sqrt{\frac{2 \cdot \Delta p}{\rho_f}} \tag{18}$$

μ–flow factor of the injector holes [-]; $\mu \approx 0.7$ according to (Orzechowski & Prywer, 1991),

Δp–pressure drop inside the sprayer [Pa],

ρ_f–fuel density [kg/m³],

a_u–factor of free-stream turbulence in the spray tip layer [-],

$$a_u = C_1 \cdot We^k \cdot Lp^l \cdot M^m \tag{19}$$

We–dimensionless Weber criteria number [-],

$$We = \frac{\rho_f \cdot w^2 \cdot d_r}{\sigma_f} \tag{20}$$

Lp–dimensionless Laplace criteria number [-],

$$Lp = \frac{\rho_f \cdot \sigma_f \cdot d_r}{\eta_f^2} \tag{21}$$

M–air to fuel density ratio [-],

$$M = \frac{\rho_g}{\rho_f} \tag{22}$$

w–relative velocity of fuel droplets inside the spray jet [m/s]; $w = w_0$,

d_r–diameter of sprayer holes [m],

σ_f–fuel surface tension [N/m],

ρ_f–fuel density [kg/m³],

ρ_g–cylinder charge density (air density) [kg/m³],

η_f–fuel absolute viscosity [kg/(m·s)],

C_1, k, l, m–experimental constants (see Table 1).

For high backpressure (M = 0.0095 - 0.028)	For low backpressure (M = 0.0014 - 0.0095)
C_1= 2.72 k= −0.21 l= 0.16 m= 1	C_1= 0.202 k= −0.21 l= 0.16 m= 0.45

Table 1. The values of experimental constants C_1, k, l, m used for calculation of spray tip penetration (Orzechowski & Prywer, 1991)

The spray tip angle, similarly to the spray tip penetration, is the function of parallel known parameters, i.e. densities of fuel and cylinder charge (ρ_f, ρ_g), fuel absolute viscosity and surface tension (η_f, σ_f), an initial velocity of fuel spray tip left the injector (w_0), diameter of sprayer holes (d_r) and time (t). Excepting an initial phase o the injection, the spray tip angle does not change, so the effect of time axis can be neglected. Further analysis is based on the following formula:

$$tg\ \alpha = C \cdot We^k \cdot Lp^l \cdot M^m \tag{23}$$

hence:

$$\alpha = arctg\ (C \cdot We^k \cdot Lp^l \cdot M^m) \tag{24}$$

where:

α–a half of spray tip angle [rad],

We, Lp, M–same numbers as for equations (20)-(22),

C, k, l, m–experimental constants (see Table 2).

For high backpressure (M = 0.0095 – 0.028)	For low backpressure (M = 0.0014 – 0.0095)
C= 0.0089 k= 0.32 l= 0.07 m= 0.5	C= 0.0028 k= 0.32 l= 0.07 m= 0.26

Table 2. The values of experimental constants C, k, l, m used for calculation of spray tip angle (Orzechowski & Prywer, 1991)

The last but not least of the analyzed parameters that is essential for this methodology is microstructure parameter of the spray jet, i.e. mean diameter of droplets. It means an equivalent average value respecting the whole spectrum of diameters of actual droplets generated by the injector. There are a few definitions of equivalent droplet size. Each of them can be described according to the general formula:

$$d_{p,q} = {}^{p-q}\!\!\sqrt{\frac{\sum n_i \cdot d_i^p}{\sum n_i \cdot d_i^q}} \tag{25}$$

where:

$d_{p,q}$-theoretical equivalent mean diameter of droplets in a spray jet [mm],

p, q-the exponents that correspond with adopted definition of droplet mean diameter [-]; the values of p and q and formula shape for various definitions is given in Table 3,

d_i-an actual diameter of droplet in spray jet [mm],

n_i-the number of droplets of the actual diameter d_i [-].

Equivalent mean diameter of droplets	p	q	Calculation formula
Arithmetic d_{10}	1	0	$d_{1,0} = \dfrac{\sum n_i \cdot d_i}{\sum n_i}$
Areal d_{20}	2	0	$d_{2,0} = \sqrt{\dfrac{\sum n_i \cdot d_i^2}{\sum n_i}}$
Areal comparative d_{21}	2	1	$d_{2,1} = \dfrac{\sum n_i \cdot d_i^2}{\sum n_i \cdot d_i}$
Volumetric d_{30}	3	0	$d_{3,0} = \sqrt[3]{\dfrac{\sum n_i \cdot d_i^3}{\sum n_i}}$
Volumetric comparative (Probert) d_3	3	1	$d_{3,1} = \sqrt{\dfrac{\sum n_i \cdot d_i^3}{\sum n_i \cdot d_i}}$
Volumetric-areal (Sauter) (also SMD - Sauter mean diameter) d_{32}	3	2	$d_{3,2} = \dfrac{\sum n_i \cdot d_i^3}{\sum n_i \cdot d_i^2}$

Table 3. The list of chosen definition formulas for calculation of mean diameter of droplets in a spray jet (Orzechowski & Prywer, 1991)

For combustion engine research area, the most usefulness definition is this one, given by Sauter formula $d_{3,2}$ (Table 3). It allows the most accurate rendering of the phenomena, where evaporation, heat and mass transfer, and combustion proceeds and is strictly crucial. Since the equation of SMD definition can be used only for research of mere injection process, comparative studies give different empirical formulas for calculations with using other parameters. For example, Hiroyasu and Katoda (Hiroyasu & Katoda, 1976) elaborated the experimental formula which is convenient to use in engine fuel injection and combustion studies. The equation, which has been consequently used by other researchers (Benson et. al., 1979; Heywood, 1988) is following:

$$d_{3,2} = A \cdot \Delta p^{-0.135} \; \rho_g^{\,0.121} \; q_{Vf}^{\,0.131} \tag{26}$$

where:

$d_{3,2}$-Sauter mean diameter [µm],

A-a constant for specific sprayer type [-]; for hole sprayer: A = 23.9,

Δp–fuel injection overpressure [MPa],

ρ_g–density of cylinder charge [kg/m^3],

q_{vf}–amount of a single fuel injection volume [mm^3].

The presented methodology for calculation of fuel spray jet parameters is based on extensive experimental studies. Thus it is expected to provide a good consistence of calculated results with experiments.

3.3 Fuel evaporation model

Fuel evaporation process is a predecessor of combustion, which begins to run just after the start of injection. The evaporation rate is the function of numerous factors, both spray surroundings and the mere fuel parameters. Nevertheless, the temperature inside the combustion chamber is the most important here. For a single fuel droplet, a relationship between size decreasing and evaporation intensity is known (Kowalewicz, 2000):

$$d_0^2 - d^2 = K \cdot t \tag{27}$$

where:

K–evaporation intensity factor that depends on temperature of surrounding where the fuel is injected [mm^2/s],

t–evaporation time [s],

d_0–initial diameter of droplet [mm],

d–diameter of droplet after the time t [mm].

An evaporation intensity factor K is the function of temperature and can be derived from experimental measurements. The equation (27) allows calculating the time of complete droplet evaporation by assuming $d = 0$. Also the total mass flux of the fuel vapor coming from a single droplet can be determined as follows:

$$\dot{m}_v = \frac{\pi \cdot K \cdot d_0 \cdot \rho_f}{6} \tag{28}$$

where:

$\dot{m}_v$–fuel vapor mass flux generated by evaporating single droplet of initial diameter d_0 [g/s],

ρ_f–fuel density [g/mm^3],

K, d_0–the same values as for equation (27).

The mass flux of fuel vapor coming from the entire spray jet depends on the numbers of droplets and their size distribution (atomization spectrum). Exact quantitative calculations are practically impossible here. Hence, the averaging equivalent values must be considered including droplet mean diameter and the number of droplets in accordance with actual fuel volume injected. From the droplet equivalent size theory we can estimate the number of droplets of Sauter mean diameter $d_{3,2}$ covered by the spray jet consisting of the liquid fuel of a volume V_f:

$$x = \frac{6 \cdot V_f}{\pi \cdot d_{3,2}^3} \tag{29}$$

where:

x–the number of droplets of Sauter mean diameter $d_{3,2}$ inside the spray jet [-],

V_f–the volume of injected and atomized fuel [mm³],

$d_{3,2}$–Sauter mean diameter (SMD) [mm].

The mass flux of fuel vapor that comes from the entire stream jet is calculated as follows:

$$\dot{M}_v = \dot{m}'_v \cdot x = \frac{\pi \cdot p_1 K \cdot d_{3,2} \cdot \rho_f}{6} \cdot \frac{6 \cdot V_f}{\pi \cdot d_{3,2}^3} = \frac{p_1 K \cdot \rho_f \cdot V_f}{d_{3,2}^2} \tag{30}$$

where:

$\dot{M}_v$–total mass flux of fuel vapor [g/s],

$\dot{m}'_v$–mass flux of fuel vapor comes from evaporation of single droplet located in a cloud of droplets [g/s],

x–the number of droplets in the stream jet [-],

p_1–factor correcting vaporization intensity of droplets (K) located in the cloud; typical value of p_1 is $p_1 \approx 0.8$-0.9,

K, d_{32}, V_f, ρ_f – the same values as for equations (27)-(29).

The value of p_1, according to (Kowalewicz & Mozer, 1977; Kucharska-Mozer, 1975; Mozer, 1976), respects the impact of cloud of droplets on the single evaporating droplet. It decreases the value of evaporation intensity factor K, because the close presence of other evaporating droplets slows down vaporization due to local temperature lowering.

Finally, it was assumed that the change of droplets size produces the same effect as the change of quantity of droplets of unchanged diameter. Then, the instantaneous fuel vaporization speed (total fuel vapor flux) can be expressed by differential equation:

$$\frac{dM_v}{d\varphi} = \frac{p_1 K\left(T_{II}(\varphi)\right) \cdot \rho_f \cdot V_f(\varphi)}{6 \cdot n \cdot d_{3,2}^2} \cdot 10^{-6} \tag{31}$$

where:

$dM_v/d\varphi$–instantaneous fuel vaporization speed [kg/deg],

φ–independent variable: the current crankshaft angle position [deg],

$K(T_{II}(\varphi))$–evaporation intensity factor as the function of the zone II temperature [mm²/s], n–crankshaft rotational speed [1/min],

$V_f(\varphi)$–instantaneous volume of liquid fuel in the stream jet [m³],

  –remaining denotations are as same as in equations (28)-(30).

Instantaneous volume of liquid fuel in the stream jet depends on fuel injection rate and fuel vaporization speed. It can be described by the following differential equation:

$$\frac{dV_f}{d\varphi} = \dot{V}_{inj}(\varphi) - \dot{V}_v(\varphi) = \dot{V}_{inj}(\varphi) - \frac{\dot{M}_v(\varphi)}{\rho_f} =$$
$$= \dot{V}_{inj}(\varphi) - \frac{p_1 K(T_{II}(\varphi)) \cdot V_f(\varphi)}{6 \cdot n \cdot d_{3,2}^2} \cdot 10^{-6}$$

(32)

where:

$dV_f/d\varphi$–change of liquid fuel volume in the stream jet [m³/deg],

$\dot{V}_{inj}(\varphi)$ –volumetric fuel injection rate [m³/deg],

  –remaining denotations are as same as in equation (31).

By resolving equations (30) and (31) that are related each to other we get the rate of evaporated fuel as the function of crankshaft rotation angle. When the calculations exceed the moment of an autoignition, combustion will start and consequntly, the equation (31) will cover the additional component describing fuel vapor loss due to its burning.

3.4 Models for formation of chemical compounds

The chemistry of combustion and formation of different compounds can be included into the overall structure of the presented model. It will be shown on the example of NO formation, where the two of reversible Zeldovich's reactions will be analyzed (Zeldovich et. al., 1947, as cited in Heywood, 1988; Kafar & Piaseczny, 1998):

$$O + N_2 \leftrightarrow NO + N$$

(33)

$$N + O_2 \leftrightarrow NO + O$$

(34)

On the base of chemical kinetic theory, the formula to calculate the NO formation rate according to the above reaction scheme is following:

$$\frac{1}{V} \cdot \frac{dn_{NO}}{dt} = 2k_1 \cdot [O] \cdot [N_2]$$

(35)

where:

V–volume of reaction zone [m³],

n_{NO}–mole number of NO [mole],

t–time [s],

k_1–kinetic constant of the first Zeldovich reaction in forward direction [m³/(mole ·s)],

$[O]$,$[N_2]$-molar concentration of O-atoms and N_2-molecules inside the reaction zone $[mole/m^3]$.

It proves that the formation rate is controlled by the first Zeldovich reaction. Atoms of oxygen come mainly from dissociation process $O_2 \leftrightarrow 2\,O$, and their concentration can be calculated as follows:

$$[O] = \left(K^c_{\,O} \cdot [O_2]\right)^{\frac{1}{2}} \tag{36}$$

where:

K^c_O-equilibrium constant of oxygen dissociation reaction referred to the concentration $[mole/m^3]$,

$[O]$,$[O_2]$-molar concentration of O-atoms and O_2-molecules inside the reaction zone $[mole/m^3]$.

Finally, the NO formation rate formula (35) takes a following shape (all denotations are as same as above):

$$\frac{1}{V} \cdot \frac{dn_{NO}}{dt} = 2k_1 \cdot K^c_O{}^{\frac{1}{2}} \cdot [O_2]^{\frac{1}{2}} \cdot [N_2] \tag{37}$$

The same formula can express a mass flux of NO in kilograms, so as to be used directly in the model differential equation system:

$$\frac{dM_{NO}}{d\varphi} = \frac{\mu_{NO} \cdot V}{6000 \cdot n} \cdot \left[2k_1 \cdot K^c_O{}^{\frac{1}{2}} \cdot [O_2]^{\frac{1}{2}} \cdot [N_2] \right] \tag{38}$$

where:

μ_{NO}-molar mass of nitric oxide $[g/mole]$; $\mu_{NO} = 30.0061$,

n-engine crankshaft speed $[rev/min]$,

-remaining denotations are as same as above.

The constants K^c_O and k_1 can be gathered from the bibliography sources (Heywood, 1988; Rychter & Teodorczyk, 1990), and are equal to:

$$k_1 = 7,6 \cdot 10^7 \cdot \exp\left(\frac{-38000}{T}\right) \left[\frac{m^3}{mole \cdot s}\right] \tag{39}$$

$$K^c_O = \frac{10^{\left[\begin{array}{c} 5+0,310805 \cdot \ln(T) - \frac{12954}{T} + \\ +1,07083 - 0,738336 \cdot 10^{-4} \cdot T + \\ +0,344645 \cdot 10^{-8} \cdot T^2 \end{array}\right]}}{\overline{R}T} \left[\frac{mole}{m^3}\right] \tag{40}$$

4. Conclusion

The mathematical model of combustion that is presented in brief in this chapter consists of a lot of phenomena. Here, the most important like the energy conversion, fuel injection and NO_X formation are presented. Many of physical and chemical events occurred in the actual engine have been omitted in the model or considered in a reduced form. It is because of impossibility in their exact mathematical representation. Surely, it influences model accuracy, but can be partially compensated by pre-calculation parametric estimation process. This way of model validation shows the disadvantage, i.e. it has to be anew performed for each engine taken under simulation. Nevertheless, the presented model can be a valuable research tool to be used for extensive studies on combustion in all types of stratified charge engines.

5. References

Benson, R.S. & Whitehouse, N.D. (1979). *Internal Combustion Engines*. Pergamon Press, ISBN 978-008-0227-20-7, Oxford

Heywood, J.B. (1988). *Internal Combustion Engines Fundamentals*. McGraw Hill Co., ISBN 0-07-028637-X, New York, NY

Hiroyasu, H. & Kadota, T. (1976). Models for Combustion and Formation of Nitric Oxide and Soot in Direct Injection Diesel Engines. SAE Technical Paper 760129, *SAE Trans*, SAE Inc., Warrendale, PA

Kafar, I. & Piaseczny, L. (1998). Mathematical Model of Toxic Compound Emissions in Exhaust Gases Produced by Marine Engines. *Journal of KONES*, Vol.5, No.1, pp. 131-141, ISSN 1231-4005, Warsaw-Gdańsk, Poland

Khan, I.M., Greeves, G. & Wang C.H.T. (1973). Factors Affecting Smoke and Gaseous Emissions from Direct Injection Engines and a Method of Calculation. SAE Technical Paper 730169, *SAE Trans*, SAE Inc., Warrendale, PA

Kowalewicz, A. & Mozer, I. (1977). Method for cylinder pressure and temperature rate calculation in compression-ignition engine based on fuel injection characteristics (in Polish: Metoda określania przebiegu ciśnień i temperatur w cylindrze silnika wysokoprężnego na podstawie charakterystyki wtrysku). *5th International Simposium on Combustion Processes*, Cracow, Poland

Kowalewicz, A. (2000). *Fundamentals of Combustion Processes* (in Polish: Podstawy procesów spalania). WNT, ISBN 83-204-2496-8, Warsaw, Poland

Kucharska-Mozer, I. (1975). *Simulation of cylinder pressure and temperature in direct injection diesel engine based on fuel injection characteristics. Part I.* (In Polish: Metoda symulacji przebiegu temperatur i ciśnień w cylindrze silnika wysokoprężnego z wtryskiem bezpośrednim na podstawie charakterystyki wtrysku). Report No.3.41.113, Institute of Aviation, Warsaw, Poland

Kuszewski, H. & Szlachta, Z. (2002). Ecological point of fuel spray jet shaping in compression-ignition engine (in Polish: Ekologiczny aspekt kształtowania własności strugi rozpylanego paliwa w silniku wysokoprężnym). *Problemy Eksploatacji*, No.1, pp. 163-173, ITE, Radom, Poland

Mozer, I. (1976). *Simulation of cylinder pressure and temperature in direct injection diesel engine based on fuel injection characteristics. Part II.* (In Polish: Metoda symulacji przebiegu temperatur i ciśnień w cylindrze silnika wysokoprężnego z wtryskiem bezpośrednim na podstawie charakterystyki wtrysku - część II: Symulacja cyfrowa). Report No.3.41.1.6, Institute of Aviation, Warsaw, Poland

Orzechowski, Z. & Prywer, J. (1991). *Spraying liquids* (in Polish: Rozpylanie cieczy). WNT, ISBN 83-204-1378-8, Warsaw, Poland

Patterson, M.A.; Kong, S.C.; Hampson, G.J. & Reitz, R.D. (1994). Modeling the Effects of Fuel Injection Characteristics on Diesel Engine Soot and NO$_X$ Emissions. SAE Technical Paper 940523, *SAE Trans*, SAE Inc., Warrendale, PA

Patterson, M.A. (1997). *Modeling the Effects of Fuel Injection Characteristics on Diesel Combustion and Emissions*. Ph.D. Thesis, University of Wisconsin-Madison

Rychter, T. & Teodorczyk, A. (1990). *Mathematical Modeling of Piston Engine Working Cycle* (in Polish: Modelowanie matematyczne roboczego cyklu silnika tłokowego). PWN, ISBN 978-830-1096-42-7, Warsaw 1990,.

Shaded, S.M.; Chiu, W.S. & Yumlu, V.S. (1973). A Preliminary Model for the Formation of Nitric Oxide in Direct Injection Diesel Engines and Its Application in Parametric Studies. SAE Technical Paper 730083, *SAE Trans*, SAE Inc., Warrendale, PA

Wajand, J.A. (1988). *Compression ignition engines* (in Polish: Silniki o zapłonie samoczynnym), WNT, ISBN 978-832-0401-68-4, Warsaw, Poland

Woś, P. (2008). *Effect of Fuel Injection Rate on Combustion and NO$_X$ Emission in Diesel Engine. Mathematical approach* (in Polish: Wpływ przebiegu wtrysku paliwa w silniku wysokoprężnym na spalanie i emisję NO$_X$). Rzeszów University of Technology, ISBN 978-83-7199-519-4, Rzeszów, Poland

Zabłocki, M. (1976). Fuel Injection and Combustion in Diesel Engines (in Polish: Wtrysk i spalanie paliwa w silnikach wysokoprężnych). WKŁ, Warsaw, Poland

4

Experimental Investigation on Premixed Combustion in a Diesel Engine with Ultra-Multihole Nozzle

Xuelong Miao, Yusheng Ju, Xianyong Wang,
Jianhai Hong and Jinbao Zheng
Wuxi Fuel Injection Equipment Research Institute
China

1. Introduction

Conventional diesel combustion exhibits the trade-off relationship between reduction of NOx and soot emissions. The strategies for reducing soot increase NOx; for example increasing injection pressure and swirl ratio reduce soot, but increase NOx in the conventional diesel combustion. High levels of exhaust gas recirculation (EGR) also reduce NOx but increase soot. It is well known that NOx emissions are largely dependent on the equivalence ratios of air-fuel mixture. NOx emissions reach maximum value when combustion occurs near stoichiometric air-fuel ratio. However, it can be lowered when mixture is over-rich or over-lean. Generally speaking, soot emission is largely produced during the diffusion combustion, but it is very low during the premixed combustion. Therefore researchers attempt to mix fuel and air as homogeneous as possible prior to ignition to achieve the premixed combustion (Kimura et al.,1999,2001; Hasegawa & Yanagihara,2003;Takeda et al.,1996; Shimazaki et al.,1999). Studies show that homogeneous charge premixed mixture low-temperature combustion can simultaneously reduce NOx and soot emissions(Nandha & Abraham,2002;Walter & Gatellier,2002; Lewander et al.,2008;Husberg et al.,2005;Lejeune et al.,2004).

The purpose of this investigation is to develop a new low-temperature premixed combustion mode in a six-cylinder commercial vehicle diesel engine using the UMH nozzle and EGR. The UMH nozzle facilitates better mixing of fuel and air prior to ignition and the resultant realization of the premixed combustion, because of its shortened the injection duration and improved atomization compared with the conventional nozzle (Miao et al.,2009). This investigation also explores the combustion characteristics of the UMH nozzle through the experiments of selected operation conditions of 1400 r/min, 0.575MPa and 1000 r/min, 0.279 by adjusting injection timing, injection pressure and EGR rate. The results showed that NOx and soot emissions of the selected operation conditions were simultaneously largely reduced.

2. Experimental apparatus

2.1 UMH nozzle structure

Figure 1 shows the schematic of the UMH nozzle (Miao et al.,2009). It consists of a needle and a body, and has the following characteristics: (1) there are two layers of injection holes in the front part of body. Any injection hole of upper layer and the corresponding injection hole of under layer are positioned in a vertical plane; (2) the injection holes cone angle (here defined as the angle of cone consisting of all the axes of injection holes on the same layer) of the under-layer holes is larger than that of the upper-layer holes ($\alpha 2 > \alpha 1$) ; and (3) it has a large enough flow area of holes such that cyclic fuel can be completely injected into the combustion chamber prior to ignition, which is a prerequisite for premixed combustion.

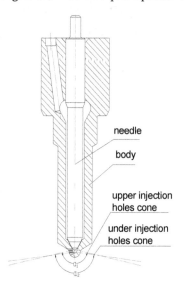

Fig. 1. Schematic of UMH nozzle

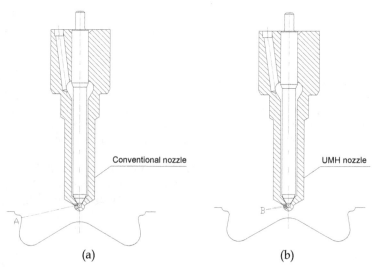

Fig. 2. Schematic comparison between the conventional and UMH nozzle (a) Conventional nozzle (b) UMH injection nozzle

Figure 2 shows the schematic comparison between the existing conventional nozzle and the UMH nozzle. There is only one layer of holes on the conventional nozzle, so sprays possibly impinge on the wall of the combustion chamber or the cylinder liner to cause high HC and CO emissions. Sprays might impinge at point A as shown in Figure 2(a). The UMH nozzle, however, has two layers of holes and the cone angle of the under-layer injection holes is larger than that of the upper-layer holes. Two sprays of upper and under-layer meet in the space of the combustion chamber, for example, at point B as shown in Figure 2(b). The results showed that the UMH nozzle exhibits shorter spray penetration than does the conventional nozzle (Miao et al.,2009). This not only avoids fuel sprays impingement on the wall of the combustion chamber or the cylinder liner, but also strengthens sprays turbulence, which promotes fuel-air mixing. Therefore, the result is a more homogeneous mixture required to perform the premixed combustion in diesel engines.

Type	Hole number	Hole diameter (mm)	Flow rate (l/min)@10MPa
Original nozzle	8	0.17	1.2
UMH nozzle	16	0.16	2

Table 1. Specifications of test nozzles

Table 1 lists the specifications of test nozzles. It can be seen that the flow rate of the UMH nozzle with smaller diameter hole is higher than that of the original nozzle by up to 67%, which helps the UMH nozzle not only shorten the injection duration and also improve fuel atomization.

2.2 Combustion experimental apparatus

The specifications of the test engine are shown in Table 2. Its operating conditions are set at 1400 r/min, 0.575MPa and 1000 r/min, 0.279MPa. The test engine is operated on the commercially available diesel fuel with the cetane number of 51 in all the test cases. The coolant temperature is set to 80±3°C. Figure 3 gives the combustion experimental apparatus. The EGR cooler and the intercooler are water-cooled, with water circulation volume and water temperature that are adjustable. The inlet air temperature after the intercooler is maintained at 40±3°C during the whole experiment. The fuel injection is performed by a high-pressure common-rail electric-controlled system on the engine. The exhaust gas emissions are measured using a HORIBA MEXA-7100 gas analyzer, smoke density (soot) measured using an AVL 415s smoke meter, and particle matter (PM) measured using an AVL 472 partial-flow particulate sampler which allows double particulate filters to be exposed. In-cylinder pressure is acquired using a KISTLER cylinder pressure sensor.

Model	CA6DF2
Type	In-line, supercharged, inter-cooled
Cylinder number-bore×stroke（mm）	6-110×115
Rated power/speed（kW /r/min）	155/2300
Maximum torque/speed（N.m/r/min）	680/1400
Minimum torque/speed（N.m/r/min）	580~620/1000
Minimum BSFC*（g/kW.h）	205

Model	CA6DF2
Combustion chamber	reentrant
Compression ratio	16.5
Ricardo swirl ratio of inlet port	2.8

*BSFC is an acronym for 'brake specific fuel consumption'.

Table 2. Specifications of the test engine

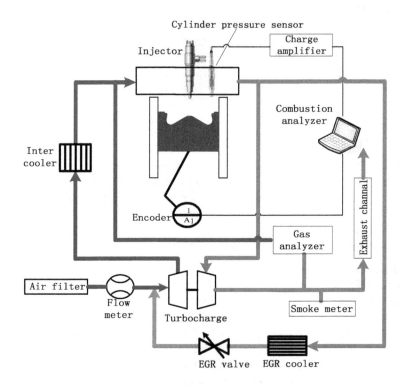

Fig. 3. Combustion experimental apparatus

EGR rate is denoted by the follow formula.

$$EGR(\%) = \frac{CO_{2,I} - CO_{2,A}}{CO_{2,E} - CO_{2,I}}$$

where, E, I, and A denote exhaust gas, inlet gas and atmosphere respectively.

3. Experimental results and discussions

The operating conditions are set at 1400 r/min, 0.575MPa, 1.18-2.29 of excess air ratio (λ) (EGR rates from 0 to 33%) (referred to as case A) and 1000 r/min, 0.29MPa, 1.68-2.92 of λ

(EGR rates from 0 to 80%) (referred to as case B). Experiments are carried out using the UMH nozzle and the original nozzle respectively. The cyclic fuel can't be completely injected into the combustion chamber before ignition because of smaller flow rate of the original nozzle, so experiment with the original nozzle can only achieve the conventional combustion. It means that this combustion can't eliminate the trade-off relationship between reduction of NOx and soot emissions, and accordingly the engine with the original nozzle is only tested in original condition. Experiments with the UMH nozzle, however, are carried out by adjusted EGR rates, injection pressures and injection timings to achieve the low-temperature premixed combustion. These optimum parameters (include EGR rate, injection pressure and injection timing) are different for case A and B to achieve the minimum values of NOx and soot emissions while keep the break specific fuel consumption (BSFC) not to be significantly deteriorated because their excess air ratios are different.

3.1 Effects of injection timing on combustion characteristics

The EGR rates of case A and B are set at 28% and 80% respectively, and the injection pressure is all 110MPa. NOx, soot, HC, CO, BSFC and cylinder pressure are measured by varying the injection timing. The results are shown in Figure 4.

It can be seen that NOx and soot emissions are simultaneously decreased by 43% and 94% respectively with retarding the injection timing from -4°ATDC to 3°ATDC in case A. For case B, NOx and soot emissions are also simultaneously decreased by 42% and84% respectively with retarding the injection timing from -4°ATDC to -1.5°ATDC, further retarding to -1°ATDC causes continuing reduction of NOx but increase of soot.

It is not difficult to understand NOx reduction with retarding the injection timing. The heat release rates at different injection timing are shown in Figure 5. The fuel injection rate curves are also plotted in Figure 5 and set at the same start point, accordingly the corresponding heat release rate curves must be shifted. In this way, it is convenient to compare the combustion characteristics, and to distinguish if the premixed combustion at different injection timing is achieved.

The premixed combustion of this investigation means combustion that occurs after the cyclic fuel completely injected into the combustion chamber. Therefore it is important the duration between the injection end point and the combustion start point (Shimazaki,2003). This duration affects combustion characteristics especially emissions, because it represents the degree of the premixed combustion. Here it is defined as the premixed degree duration denoted by τ_{pmix}. The cyclic fuel has not been completely injected into the combustion chamber prior to ignition when τ_{pmix} is less than zero, it means the complete premixed combustion can't be achieved, still belongs to the conventional combustion. However, the cyclic fuel has just been completely injected into the combustion chamber prior to ignition when τ_{pmix} is equal to zero, but it is short for fuel and air to completely mix, which can not form homogeneous mixture. The homogeneity of mixture tends to improve with the increase of τ_{pmix}, and accordingly soot and NOx emissions tend to decrease simultaneously when high levels of EGR were used. So τ_{pmix} is a very important parameter to help compare between the premixed combustion and the conventional combustion.

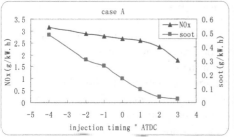

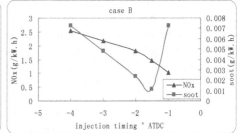

a) Effects of injection timing on NOx and soot emissions

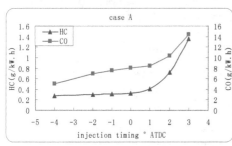

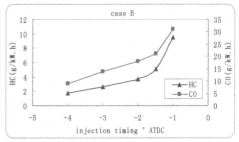

b) Effects of injection timing on HC and CO emissions

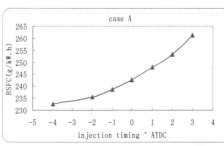

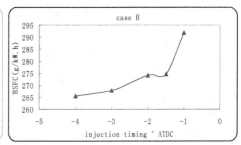

c) Effects of injection timing on BSFC

Fig. 4. Effects of injection timing on engine performance

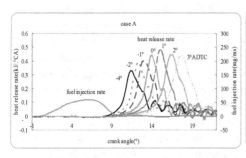

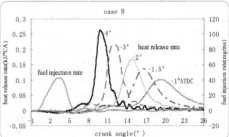

Fig. 5. Heat release rates at different injection timing

The premixed degree duration τ_{pmix} at different injection timings is shown in Table 3. It can be seen that the premixed combustion has not been achieved at -4°ATDC of the injection timing in case A. The premixed combustion has just been achieved at -1°ATDC of the injection timing while τ_{pmix} is equal to zero. The τ_{pmix} is equal to 0.3°CA at 3°ATDC of the injection timing, which means that it is longer for fuel and air to mix prior to ignition. The longer τ_{pmix} is, the more homogeneous mixture is. In this way, it is possible to achieve the homogeneous charge combustion, eventually soot and NOx emissions are simultaneously reduced to very low when high levels of EGR were used. This is different with the conventional combustion. For case B, the premixed combustion has already been achieved at -4°ATDC of the injection timing while τ_{pmix} is equal to 3.43°CA, so soot is low. The τ_{pmix} is equal to 3.9°CA with retarding the injection timing to -2°ATDC, therefore soot is already very low. But it can be seen from Figure 5 that the combustion rate is very low with further retarding the injection timing to -1°ATDC, which is not beneficial to complete combustion, and accordingly causes increase of soot.

case	Injection start θ1	Injection end θ2	Combustion start θ3	Ignition delay =θ3-θ1	T_{pmix} =θ3-θ2
	°ATDC	°ATDC	°ATDC	°CA	°CA
A	-4	6	4.3	8.3	-1.7
	-2	8	6.6	8.6	-1.4
	-1	9	7.8	8.8	-1.2
	0	10	9.3	9.3	-0.7
	1	11	10.7	9.7	-0.3
	2	12	12.1	10.1	0.1
	3	13	13.3	10.3	0.3
B	-4	1.6	5.03	9.03	3.43
	-3	2.6	6.18	9.18	3.58
	-2	3.6	7.5	9.5	3.9
	-1.5	4.1	8.2	9.7	4.1
	-1	4.6	8.9	9.9	4.3

Table 3. Premixed degree duration τ_{pmix} at different injection timing

HC, CO and BSFC have a slight change until 1°ATDC of the injection timing, but the further retarding injection timing causes deterioration of these performances in case A. These performances are significantly worsened after 1°ATDC of the injection timing because the combustion period is far away from the top dead center (TDC). For case B due to higher EGR rate, HC and CO emissions are swiftly increased with retarding the injection timing to after -1.5°ATDC because of slowing combustion rate, and BSFC is also worsened due to far away from TDC of the combustion period.

3.2 Effects of EGR rate on combustion characteristics

The injection timing of case A and B are set at 2°ATDC and -1.5°ATDC respectively, injection pressure is all 110MPa. NOx, soot, HC, CO, BSFC and cylinder pressure are measured by varying EGR rates. The results are shown in Figure 6.

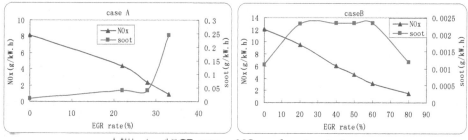

a) Effects of EGR rate on NOx and soot emissions

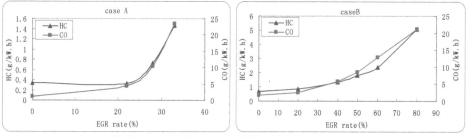

b) Effects of EGR rate on HC and CO emissions

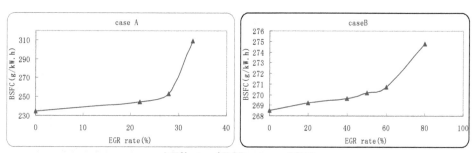

c) Effects of EGR rate on BSFC

Fig. 6. Effects of EGR rate on engine performance

It can be seen from Figure 6 that NOx is linearly decreased with the increase of the EGR rate. For case A, NOx is decreased by 71% and 89% respectively when the EGR rate is from zero to 28% and 33%. Soot is almost unchanged when the EGR rate is less than 28%, but further increasing the EGR rate causes swift increase of soot emission. For case B, NOx is decreased by 88% when the EGR rate is from zero to 80%, but soot has a complicated tendency. Firstly soot has a slight change when the EGR rate is low, then reaches the maximum value when the EGR rate is 20%, however soot begins to decrease swiftly with further increase of the EGR rate.

Figure 7 shows the heat release rates and Table 4 shows the premixed degree duration τ - pmix at different EGR rates. The reduction of oxygen concentration with EGR causes NOx decrease and soot increase, but on the other hand, longer τ pmix due to EGR causes soot decrease. Therefore effects of EGR on soot emission are as follows. Lower EGR rates don't cause significant change of soot. But soot is deteriorated when the EGR rate is more than 28%

in case A due to lower excess air ratio (lower oxygen concentration). Longer τ_{pmix}, however, dominates combustion process compared to oxygen concentration decrease when the EGR rate is high in case B due to higher excess air ratio. There is much time for fuel and air to mix prior to ignition with the aid of EGR. This is beneficial to the formation of a homogeneous mixture. Therefore soot has been greatly decreased again when EGR rate is high in case B.

HC, CO and BSFC have a slight change when the EGR rate is less than 22%, then begin to increase with further increase the EGR rate, especially worsen when the EGR rate reaches 33% in case A. For case B, however, HC, CO and BSFC have a slight change when EGR rate is less than 30%, then begin to worsen with further increase of the EGR rate. This is because of an incomplete combustion causing with the EGR rate increase.

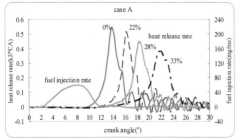 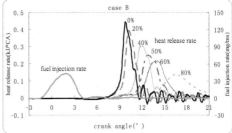

Fig. 7. Heat release rates at different EGR rates

case	EGR rate	Injection start θ1	Injection end θ2	Combustion start θ3	Ignition delay =θ3-θ1	τ_{pmix} =θ3-θ2
	%	°ATDC	°ATDC	°ATDC	°CA	°CA
A	0	2	12	9.7	7.7	-2.3
	22	2	12	11.6	9.6	-0.4
	28	2	12	12.1	10.1	0.1
	33	2	12	12.6	10.6	0.6
B	0	-1.5	4.1	6.94	8.44	2.84
	20	-1.5	4.1	6.95	8.45	2.85
	40	-1.5	4.1	7.52	9.02	3.42
	50	-1.5	4.1	7.73	9.23	3.63
	60	-1.5	4.1	7.75	9.25	3.65

Table 4. Premixed degree duration τ_{pmix} at different EGR rates

3.3 Effects of injection pressure (rail pressure) on combustion characteristics

The injection timing of case A and B are set at 2°ATDC and -1.5°ATDC respectively, EGR rates are set at 28% and 80% respectively. NOx, soot, HC, CO, BSFC and cylinder pressure are measured by varying the injection pressure. The results are shown in Figure 8.

It can be seen that soot, HC, CO and BSFC have some certain reduction except NOx with the injection pressure increase in case A. For case B, these changes are almost the same as case A when the injection pressure is less than 110MPa. But these performances are deteriorated with the injection pressure further increase. This is maybe due to the spray wall-impingement with the injection pressure further increase.

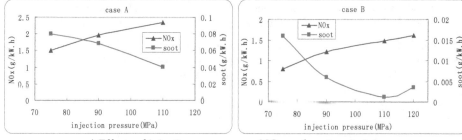

a) Effects of injection pressure on NOx and soot emissions

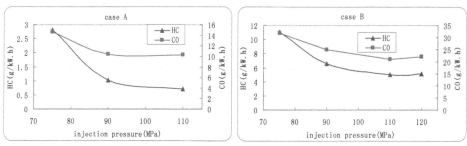

b) Effects of injection pressure on HC and CO emissions

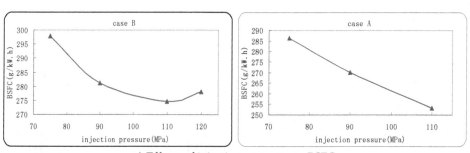

c) Effects of injection pressure on BSFC

Fig. 8. Effects of injection pressure on engine performance

Figure 9 shows the heat release rates and Table 5 shows the premixed degree duration τ_{pmix} at different injection pressures. It can be seen that combustion advance and heat release peak increase with the injection pressure increase, and thus NOx is increased. But the advance of injection end point leads to longer τ_{pmix} with the injection pressure increase. In case A, τ_{pmix} increases from -0.5°CA to 0.1°CA with the injection pressure increase from 75MPa to 110MPa, so soot is decreased. Additionally, BSFC can be improved due to the combustion advance with the injection pressure increase. Higher injection pressure, of course, can improve mixing of fuel and air, finally improve combustion. Therefore HC and CO emissions are also decreased. In case B, although τ_{pmix} keeps almost unchanged with the injection pressure increase from 90MPa to 110MPa, but the higher injection pressure can improve mixing of fuel and air, so soot can be significantly reduced. Over spray penetration, however, leads to bad performance when the injection pressure reaches 120MPa.

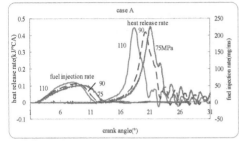 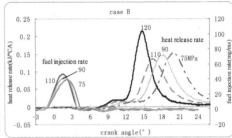

Fig. 9. Heat release rates at different injection pressure

case	Injection start θ1	Injection pressure	Injection end θ2	Combustion start θ3	Ignition delay =θ3-θ1	τ_{pmix} =θ3-θ2
	°ATDC	MPa	°ATDC	°ATDC	°CA	°CA
A	2	110	12	12.1	10.1	0.1
		90	12.5	12.8	10.8	0.3
		75	13.4	12.9	10.9	-0.5
B	-1.5	120	4	7.72	9.22	3.72
		110	4.1	8.2	9.7	4.1
		90	4.2	8.92	9.22	4.72
		75	4.7	9.5	11	4.8

Table 5. Premixed degree duration τ_{pmix} at different injection pressure

3.4 Combustion characteristics comparison between UMH nozzle and original nozzle

Figure 10 shows that the relationship of NOx and soot emissions is compared between the UMH nozzle and the original nozzle (original engine). Because engine with the original nozzle can't achieve the complete premixed combustion and still belongs to the conventional combustion, which trade-off relationship between reduction of NOx and soot emissions can't be overcome. So the engine with the original nozzle is only tested in the original condition, this result is used to compare with that of the UMH nozzle. The engine with the UMH nozzle, however, can achieve the premixed combustion to obtain simultaneous reduction of NOx and soot emissions. Therefore it is tested by adjusted the EGR rate, injection pressure and injection timing to obtain a series of values of NOx and soot emissions. It can be seen that NOx is high but soot very low without EGR, and then NOx is swiftly decreased with the EGR rate increase. In sum, for case A, NOx and soot emissions can simultaneously achieve minimum values when EGR rate is 28%, and combing suitable injection pressure and injection timing. For case B, NOx is very low when EGR rate is 80%, and soot is also simultaneously reduced to very low value combing suitable injection pressure and injection timing. Table 6 shows combustion characteristics comparison between the UMH nozzle and the original nozzle. NOx and soot emissions of the UMH nozzle are reduced by 68.2% and 20% respectively than that of the original nozzle in case A, but BSFC increases by 10.4%. For case B, NOx and soot emissions of the UMH nozzle are reduced by 78.1% and 76.2% respectively than that of the original nozzle, but BSFC only increases by 8.7%. This is because the excess air ratio of case B is higher than that of case A. Therefore case B can use higher EGR rate than case A, which leads to the achievement of much more lean-homogeneous charge combustion.

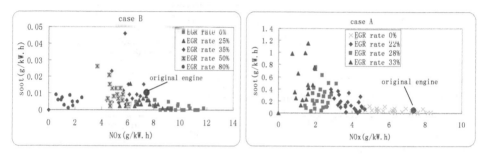

Fig. 10. NOx − soot emissions comparison between UMH and original nozzle

case	Nozzle type	EGR rate	excess air ratio	Injection pressure	Injection start	NOx	HC	CO	soot	BSFC	τ_pmix
		%	λ	MPa	°ATDC			g/kW.h			°CA
A	Original nozzle	0	2.25	75	-2	7.35	0.54	1.38	0.05	235	-2
	UMH nozzle	28	1.49	110	2	2.34	0.73	10.33	0.04	259.4	0.1
B	Original nozzle	0	3.5	50	0	6.73	1.52	7.69	0.005	252.6	-4.6
	UMH nozzle	80	1.68	110	-2	1.47	5.08	21.15	0.001	274.8	4.1

Table 6. Combustion characteristics comparison between UMH and original nozzle

4. Conclusions

1. The UMH nozzle has a large flow area of holes, which is beneficial to homogeneous mixture preparation prior to ignition. The better premixed combustion can be achieved combing the UMH nozzle with EGR, suitable injection pressure and injection timing to obtain simultaneous reduction of NOx and soot emissions in a diesel engine.

2. For case A, NOx and soot emissions of the UMH nozzle compared with the original nozzle are simultaneously reduced by 68.2% and 20% respectively when the EGR rate is

28% (excess air ratio is 1.49), the injection pressure is 110MPa and the injection timing is 2°ATDC. For case B, however, NOx and soot emissions of the UMH nozzle compared with the original nozzle are simultaneously reduced by 78.1% and 76.2% respectively when the EGR rate is 80% (excess air ratio is 1.68), the injection pressure is 110MPa and the injection timing is -1.5°ATDC.

3. Case B can use higher EGR rate to achieve much more lean-homogeneous charge premixed combustion because the excess air ratio of case B is higher than that of case A. Therefore it can obtain remarkably simultaneous reduction of NOx and soot emissions and slight increase of BSFC by 8.7% compared with the original engine.

5. Acknowledgements

This study was financially supported by Wuxi Fuel Injection Equipment Research Institute. The authors wish to express their gratitude to the Executive of Wuxi Fuel Injection Equipment Research Institute.

6. References

Hasegawa,R.& Yanagihara,H.(2003). *HCCI combustion in diesel engine.* SAE Paper 2003-01-0745.

Husberg,T; Gjirja,S.& Denbratt,I.(2005). *Fuel Equivalence Ratio and EGR Impact on Premixed Combustion Rate and Emission Output, on a Heavy-Duty Diesel Engine.* SAE Paper 2005-24-046.

Kimura,S.; Aoki,O.& Ogawa,H.(1999). *New Combustion Concept for Ultra-Clean and High-Efficiency Small Di Diesel Engines.* SAE Paper 1999-01-3681,

Kimura,S.; Aoki,O.& Kitahara,Y.(2001). *Ultra-clean combustion technology combing low-temperature and premixed concept to meet future emission standards.* SAE Paper 2001-01-0200.

Lewander,C.,M.; Ekholm,K.& Johansson,B.(2008). *Investigation of the Combustion Characteristics with Focus on Partially Premixed Combustion in a Heavy Duty Engine.* SAE Paper 2008-01-1658.

Lejeune,M.; Lortet,D& Benajes, J.(2004).*Potential of Premixed Combustion With Flash Late Injection on a Heavy-Duty Diesel Engine.* SAE Paper 2004-01-1906.

Nandha,K.,P.& Abraham,J.(2002). *Dependence of fuel-air mixing characteristics on injection timing in early-injection diesel engine.* SAE Paper 2002-01-0944.

Shimazaki,N; Akagawa,H& Tsujimura K.(1999). *Experimental study of premixed lean diesel combustion.* SAE Paper 1999-01-0181.

Shimazaki,N; Tsurushima,T.& Nishimura,T.(2003). *Dual Mode Combustion Concept With Premixed Diesel Combustion by Direct Injection Near Top Dead Center.* SAE Paper 2003-01-0742.

Takeda,Y.; Keiichi, N.& Keiichi,N. (1996). *Emission characteristics of premixed lean diesel combustion with extremely early stage fuel injection.* SAE Paper 961163.

Walter,B.& Gatellier,B. (2002). *Development of high power NADITM concept using dual mode diesel combustion to achieve zero NOx and particulate emissions.* SAE Paper 2002-01-1744.

Xuelong,Miao; Xinqi,Qiao& Xianyong,Wang(2009). *Study on Ultramultihole Nozzle Fuel Injection and Diesel Combustion,Energy and Fuels*, vol.23,no.2, pp.740-743.

Section 3

Numerical Studies
on Injection Process Phenomena

5

Influence of Nozzle Orifice Geometry and Fuel Properties on Flow and Cavitation Characteristics of a Diesel Injector

Sibendu Som[1], Douglas E. Longman[1],
Anita I. Ramirez[2] and Suresh Aggarwal[2]
[1]Argonne National Laboratory,
[2]University of Illinois at Chicago,
USA

1. Introduction

Cavitation refers to the formation of bubbles in a liquid flow leading to a two-phase mixture of liquid and vapor/gas, when the local pressure drops below the vapor pressure of the fluid. Fundamentally, the liquid to vapor transition can occur by heating the fluid at a constant pressure, known as boiling, or by decreasing the pressure at a constant temperature, which is known as cavitation. Since vapor density is at least two orders of magnitude smaller than that of liquid, the phase transition is assumed to be an isothermal process. Modern diesel engines are designed to operate at elevated injection pressures corresponding to high injection velocities. The rapid acceleration of fluid in spray nozzles often leads to flow separation and pockets of low static pressure, prompting cavitation. Therefore, in a diesel injector nozzle, high pressure gradients and shear stresses can lead to cavitation, or the formation of bubbles.

Cavitation, in diesel fuel injectors can be beneficial to the development of the fuel spray, since the primary break-up and subsequent atomization of the liquid fuel jet can be enhanced. Primary breakup is believed to occur in the region very close to the nozzle tip as a result of turbulence, aerodynamics, and inherent instability caused by the cavitation patterns inside the injector nozzle orifices. In addition, cavitation increases the liquid velocity at the nozzle exit due to the reduced exit area available for the liquid. Cavitation patterns extend from their starting point around the nozzle orifice inlet to the exit where they influence the formation of the emerging spray. The improved spray development is believed to lead to more complete combustion process, lower fuel consumption, and reduced exhaust gas and particulate emissions. However, cavitation can decrease the flow efficiency (discharge coefficient) due to its affect on the exiting jet. Also, imploding cavitation bubbles inside the orifice can cause material erosion thus decreasing the life and performance of the injector. Clearly an optimum amount of cavitation is desirable and it is important to understand the sources and amount of cavitation for more efficient nozzle designs.

The flow inside the injector is controlled by dynamic factors (injection pressure, needle lift, etc.) and geometrical factors (orifice conicity, hydrogrinding, etc.). The effects of dynamic factors on the injector flow, spray combustion, and emissions have been investigated by various researchers including (Mulemane, 2004; Som, 2009a, 2010a; Payri, 2009). There have also been experimental studies concerning the effects of nozzle orifice geometry on global injection and spray behavior (Bae, 2002; Blessing, 2003; Benajes, 2004; Han, 2002; Hountalas, 2005; Payri, 2004, 2005, 2008; Som, 2009c). The literature review indicates that while the effect of orifice geometry on the injector flow and spray processes has been examined to some extent, its influence on engine combustion and emissions is not well established (Som, 2010b, 2011). To the best of our knowledge, the influence of nozzle geometry on spray and combustion characteristics has also not been studied numerically, mainly due to the complicated nature of flow processes associated. These form a major motivation for the present study i.e., to examine the effects of nozzle orifice geometry on inner nozzle flow under diesel engine conditions. With increasingly stricter emission regulations and greater demand on fuel economy, the injector perhaps has become the most critical component of modern diesel engines. Consequently, it is important to characterize the effects of orifice geometry on injection, atomization and combustion behavior, especially as the orifice size keeps getting smaller and the injection pressure higher. In order to achieve the proposed objectives, we first examine the effects of orifice geometry on the injector flow, including the cavitation and turbulence generated inside the nozzle.

Biofuels are an important part of our country's plan to develop diverse sources of clean and renewable energy. These alternative fuels can help increase our national fuel security through renewable fuel development while simultaneously reducing emissions from the transportation sector. Biodiesel is a particularly promising biofuel due to its compatibility with the current fuel infrastructure geared toward compression-ignition engines. Using biodiesel as a blending agent can prolong the use of petrodiesel. Biodiesel is also easily produced from domestic renewable resources such as soy, rape-seed, algae, animal fats, and waste oils. Our literature search (Som, 2010b) identified relatively few studies dealing with the injection and spray characteristics of biodiesel fuels. Since there are significant differences in the thermo-transport properties of petrodiesel and biodiesel fuels, the injection and spray characteristics of biodiesel can be expected to differ from those of petrodiesel. For instance, due to differences in vapor pressure, surface tension, and viscosity, the cavitation and turbulence characteristics of biodiesel and diesel fuels inside the injector may be significantly different. The injector flow characteristics determine the boundary conditions at the injector orifice exit, including the rate of injection (ROI) profile as well as the cavitation and turbulence levels; this can have a significant influence on the atomization and spray characteristics, and consequently on engine performance. Som et al. (Som, 2010b) compared the injection and spray characteristics of diesel and biodiesel (from soy-based feedstock) using an integrated modeling approach. This modeling approach accounts for the influence in nozzle flow effects such as cavitation and turbulence (Som, 2010a) on spray-combustion development using the recently developed Kelvin Helmholtz-Aerodynamic Cavitation Turbulence (KH-ACT) primary breakup model (Som, 2009b, 2010c). Another objective of the current study is to demonstrate a framework within which boundary conditions for spray and combustion modeling for different orifice shapes and alternate fuels of interest can be available from high-fidelity nozzle flow simulations.

2. Computational model

The commercial CFD software FLUENT v6.3 was used to perform the numerical simulation of flow inside the nozzle. FLUENT employs a mixture based model as proposed by Singhal et al. (Singhal, 2002). The two-phase model considers a mixture comprising of liquid fuel, vapor, and a non-condensable gas. While the gas is compressible, the liquid and vapor are considered incompressible. In addition, a no-slip condition between the liquid and vapor phases is assumed. Then the mixture properties are computed by using the Reynolds–Averaged continuity and momentum equations (Som, 2009a).

$$\frac{\partial u_j}{\partial x_j} = 0 \tag{1}$$

$$\rho \frac{\partial u_i u_j}{\partial x_j} = -\frac{\partial P}{\partial x_i} + \frac{\partial \tau_{ij}}{\partial x_j} \tag{2}$$

where $\tau_{ij} = (\mu + \mu_t)\left\{\frac{\partial u_i}{\partial x_j} + \frac{\partial u_j}{\partial x_i}\right\}$, $\mu_t = C_\mu \rho \frac{k^2}{\epsilon}$ is the turbulent viscosity

In order to account for large pressure gradients, the realizable $k-\epsilon$ turbulence model is incorporated along with the non-equilibrium wall functions.

$$\frac{\partial \rho u_j k}{\partial x_j} = \frac{\partial}{\partial x_j}\left[\left(\mu + \frac{\mu_t}{\sigma_k}\right)\frac{\partial k}{\partial x_j}\right] + P - \rho \epsilon \tag{3}$$

where P (production of turbulent kinetic energy) $= \mu_t \frac{\partial u_i}{\partial x_j}\left[\frac{\partial u_i}{\partial x_j} + \frac{\partial u_j}{\partial x_i}\right] - \frac{2}{3}\frac{\partial u_i}{\partial x_i}\left\{\rho k + \mu_t \frac{\partial u_k}{\partial x_k}\right\}$ (4)

$$\frac{\partial \rho u_j \epsilon}{\partial x_j} = \frac{\partial}{\partial x_j}\left[\left(\mu + \frac{\mu_t}{\sigma_k}\right)\frac{\partial \epsilon}{\partial x_j}\right] + \frac{\epsilon}{k}\left[c_1 P - c_2 \rho \epsilon + c_3 \rho k \frac{\partial u_k}{\partial x_k}\right]$$

The turbulent viscosity is modeled for the whole mixture. The mixture density and viscosity are calculated using the following equations:

$$\rho = \alpha_v \rho_v + (1 - \alpha_v - \alpha_g)\rho_l + \alpha_g \rho_g \tag{5}$$

$$\mu = \alpha_v \mu_v + (1 - \alpha_v - \alpha_g)\mu_l + \alpha_g \mu_g \tag{6}$$

where ρ and μ are the mixture density and viscosity respectively, and the subscripts v, l, g represent the vapor, liquid, and gas respectively. The mass (f) and volume fractions (α) are related as:

$$\alpha_v = f_v \frac{\rho}{\rho_v}, \ \alpha_l = f_l \frac{\rho}{\rho_l}, \text{ and } \alpha_g = f_g \frac{\rho}{\rho_g} \tag{7}$$

Then the mixture density can be expressed as:

$$\frac{1}{\rho} = \frac{f_v}{\rho_v} + \frac{f_g}{\rho_g} + \frac{1 - f_v - f_g}{\rho_l} \tag{8}$$

The vapor transport equation governing the vapor mass fraction is as follows:

$$\rho \frac{\partial u_j f_v}{\partial x_j} = \frac{\partial}{\partial x_j}\left(\Gamma \frac{\partial f_v}{\partial x_j} \right) + R_e - R_c \tag{9}$$

where u_i is the velocity component in a given direction (i=1,2,3), Γ is the effective diffusion coefficient, and R_e, R_c are the vapor generation and condensation rate terms (Brennen, 1995) computed as:

$$R_e = C_e \frac{\sqrt{k}}{\sigma} \rho_l \rho_v \left(1 - f_v - f_g\right)\sqrt{\frac{2(P_v - P)}{3\rho_l}}$$

$$R_c = C_c \frac{\sqrt{k}}{\sigma} \rho_l \rho_v f_v \sqrt{\frac{2(P - P_v)}{3\rho_l}} \tag{10}$$

where σ and P_v are the surface tension and vapor pressure of the fluid respectively, and k and P are the local turbulent kinetic energy and static pressure respectively. An underlying assumption here is that the phenomenon of cavitation inception (bubble creation) is the same as that of bubble condensation or collapse. Turbulence induced pressure fluctuations are accounted for by changing the phase-change threshold pressure at a specified temperature (P_{sat}) as:

$$P_v = P_{sat} + \frac{P_{turb}}{2} \quad \text{where,} \quad P_{turb} = 0.39\rho k \tag{11}$$

The source and sink terms in equation (10) are obtained from the simplified solution of the Rayleigh-Plesset equation (Brennen, 1995). No-slip boundary conditions at the walls and symmetry boundary condition at the center line are employed for the HEUI 315-B injector simulations (cf. Figure 3a).

3. Results and discussion

This section will first present a new improved criterion for cavitation inception for production injector nozzles. This new criterion will provide a tool for assessing cavitation under turbulent regimes typical in diesel injector nozzles. The influence of nozzle orifice geometry on in-nozzle flow development will be presented next. The influence of fuel properties such as density, viscosity, surface tension, and vapor pressure on nozzle flow characteristics will be presented. Cavitation and turbulence generated inside the nozzle due to geometry and fuel changes will also be quantified.

3.1 An Improved criterion for cavitation inception

According to the traditional criterion, cavitation occurs when the local pressure drops below the vapor pressure of the fuel at a given temperature i.e., when $-p + p_v > 0$. This criterion can be represented in terms of a cavitation index (K) as:

$$K_{Classical} = \frac{p - p_b}{p_b - p_v} < -1 \Rightarrow Cavitating \tag{12}$$

where p, p_b, p_v are the local pressure, back pressure, and vapor pressure, respectively. This criterion has been extensively used in the cavitation modeling community. However, Winer and Bair (Winer, 1987) and Joseph (Joseph, 1998) independently proposed that the important parameter for cavitation is the total stress that includes both the pressure and normal viscous stress. This was consistent with the cavitation experiments in creeping shear flow reported by Kottke et al. (Kottke, 2005), who observed the appearance of cavitation bubbles at pressures much higher than vapor pressure. Following an approach proposed by Joseph (Joseph, 1998) and Dabiri et al. (Dabiri, 2007), a new criterion based on the principal stresses was derived and implemented. The formulation for the new criterion is summarized below.

Maximum tension criterion: $-p - 2\mu S_{11} + p_v > 0$

Minimum tension criterion: $-p + 2\mu S_{11} + p_v > 0$

The new criteria can be expressed in terms of the modified cavitation index as:

$$K_{max} = \frac{p + 2\mu S_{11} - p_b}{p_b - p_v} < -1 \Rightarrow Cavitating \tag{13}$$

$$K_{min} = \frac{p - 2\mu S_{11} - p_b}{p_b - p_v} < -1 \Rightarrow Cavitating \tag{14}$$

where the strain rate S_{11} is computed as:

$$S_{11} = \sqrt{\left(\frac{\partial u}{\partial x}\right)^2 + \left(\frac{\partial u}{\partial y} + \frac{\partial v}{\partial x}\right)^2} \tag{15}$$

where u, v are the velocities in x, y direction respectively.

Under realistic Diesel engine conditions where flow inside the nozzle is turbulent, turbulent stresses prevail over laminar stresses. Accounting for the effect of turbulent viscosity the new criteria is further modified as:

$$K_{max\text{-}turb} = \frac{p + 2(\mu + \mu_t)S_{11} - p_b}{p_b - p_v} < -1 \Rightarrow Cavitating \tag{16}$$

$$K_{min\text{-}turb} = \frac{p - 2(\mu + \mu_t)S_{11} - p_b}{p_b - p_v} < -1 \Rightarrow Cavitating \tag{17}$$

The experimental data from Winklhofer et al. (Winklhofer, 2001) was used for a comprehensive model validation. These experiments were conducted in a transparent, quasi-2-D geometry wherein the back pressure was varied to achieve different mass flow rates. To the best of our knowledge this experimental data-set is most comprehensive in terms of two phase information and inner nozzle flow properties.

Figure 1 presents the measured cavitation contour at injection and back pressures of 100 and 40 bar respectively, owing to a Reynolds number of 16,000 approximately. It is clearly seen from the marked red line that there is significant amount of cavitation at the orifice inlet. These cavitation contours extend to certain distance inside the orifice. The vapor fraction contour shows no cavitation (blue represents pure liquid). The classical criterion which basically is another way of representing the predicted vapor fraction contour also captures the same trend, i.e., hardly any cavitation is observed. The laminar criteria shows cavitation inception, however, no advection of the fuel vapor into the orifice is observed. The turbulent criteria seems to capture more cavitation with $C_t = 2$ agreeing better with experimental data that all the other criteria.

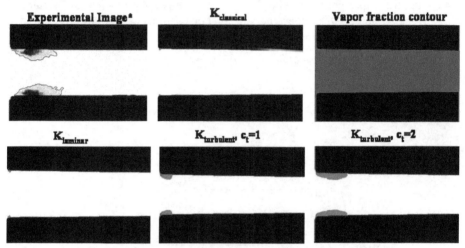

Fig. 1. Comparison between the measured (Winklhofer, 2001), predicted vapor fraction contours, and cavitation inception regions predicted by different cavitation criteria. The injection and back pressures are 100bar and 40bar respectively.

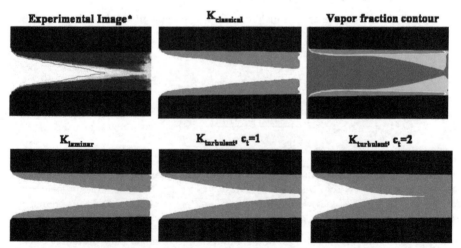

Fig. 2. Comparison between the measured (Winklhofer, 2001), predicted vapor fraction contours, and cavitation inception regions predicted by different cavitation criteria. The injection and back pressures are 100bar and 20bar respectively.

Figure 2 presents the measured cavitation contour at injection and back pressures of 100 and 20 bar respectively, owing to a Reynolds number of 18,000 approximately. It is clearly seen from the marked red line in the experimental image (Winklhofer, 2001) that there is a significant amount of cavitation at the top and bottom of orifice inlet. These cavitation contours are symmetric in nature and are advected by the flow to reach the nozzle orifice exit.

The cavitation contours join near the orifice exit thus the exit is completely covered by fuel vapor. The predicted vapor fraction contour also shows significant amount of cavitation represented by the fuel vapor contour (in red). However, there is still a significant amount of liquid fuel (in blue) present at the orifice exit and the vapor fraction contours do not join together as was the case in experiments. The classical criterion which basically is another way of representing the predicted vapor fraction also captures the trend as the vapor fraction contour. The laminar criterion predicts marginal improvement to the classical criterion. This is expected since for high Reynolds number flows the difference between these criteria was observed to diminish (Padrino, 2007). Increase in Reynolds number results in an increase in turbulence levels inside the orifice. Thus the turbulent stress criterion is seen to improve the predictions of vapor fraction contours significantly. All the experimentally observed characteristics are captured by the turbulent stress criterion i.e., the vapor contours from top and bottom of the orifice are seen to merge together resulting in pure vapor at orifice exit. The turbulent criterion seems to capture more cavitation with C_t =2 agreeing better with experimental data than all the other criteria.

The simulations using the new cavitation criterion show significant improvement in prediction of cavitation contours especially in the turbulent regime under realistic injection conditions. Future studies will focus on performing such studies in realistic geometries of interest characteristerized by three dimensional flow features. Winklhofer et al. (Winklhofer, 2001) experiments, although performed under realistic injection conditions, do not capture the 3D effects which are essential to flow development.

3.2 Effect of nozzle orifice geometry on inner nozzle flow development

This section will focus on capturing the influence of nozzle orifice geometry on in-nozzle flow development such as cavitation and turbulence in addition to flow variables such as velocity, discharge coefficient etc. The base nozzle orifice geometry which is cylindrical and non-hydroground will be presented first. The single orifice simulated for the full-production, mini-sac nozzle used in the present study is shown in Figure 3. The nozzle has six cylindrical holes with diameter of 169 μm at an included angle of 126°. The discharge coefficient (C_d), velocity coefficient (C_v) and area contraction coefficient (C_a), used to characterize the nozzle flow, are described below. The discharge coefficient (C_d) is calculated from:

$$C_d = \frac{\dot{M}_{actual}}{\dot{M}_{th}} = \frac{\dot{M}_{actual}}{A_{th}\sqrt{2 * \rho_f * \Delta P}} \tag{18}$$

where $\dot{M}_{actual}$ is the mass flow rate measured by the rate of injection (ROI) meter (Bosch, 1966), or calculated from FLUENT simulations, A_{th} is the nozzle exit area, and M_{th} is the theoretical mass flow rate. The three coefficients are related as (Naber , 1996):

$$C_d = C_v * C_a \tag{19}$$

Here the area contraction coefficient is defined as:

$$C_a = \frac{A_{effective}}{A_{th}} \qquad (20)$$

where $A_{effective}$ represents the area occupied by the liquid fuel. C_a is an important parameter to characterize cavitation, as it is directly influenced by the amount of vapor present at the nozzle exit. The Reynolds number is calculated from:

$$Re = \frac{V_{th}D_{th}\rho_{fuel}}{\mu_{fuel}} \qquad (21)$$

where D_{th} is the nozzle exit diameter.

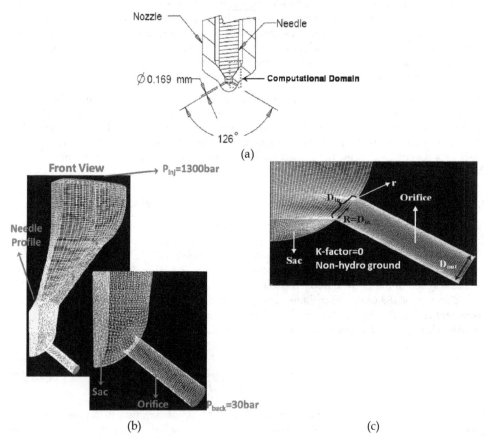

(a)

(b)   (c)

Fig. 3. (a) Injector nozzle geometry along with the computational domain. (b) The 3-D grid generated, specifically zooming in on the sac and orifice regions. (c) Zoomed 2-D view of the orifice and sac regions.

The influence of nozzle orifice geometry is characterized by comparing the in-nozzle flow characteristics of the base nozzle against a hydroground nozzle. The hydroground nozzle has the same nominal dimensions as the base nozzle except the hydrogrounding resulting in a small inlet radius of curvature. The essential features of the nozzle orifices simulated are shown in Table 1.

Geometrical Characteristics	Base Nozzle	Hydroground Nozzle
D_{in} (μm)	169	169
D_{out} (μm)	169	169
K_{factor}	0	0
r/R	0	0.014
L/D	4.2	4.2

Table 1. Geometrical Characteristics of nozzle orifice simulated.

Figure 4 presents vapor fraction contours for the base and hydroground nozzles at P_{in}=1300bar, P_b=30bar, and full needle open position. Simulations were performed for diesel fuel (properties shown in Table 2). The 3D view of the cavitation contours shows that vapor generation only occurs at the orifice inlet for both the orifices. For the base nozzle these cavitation contours are advected by the flow to reach the orifice exit. Consequently, the computed area coefficient (C_a) was found to be 0.96 for this case. A smoother orifice inlet (i.e., r/R=0.014) clearly leads to a decrease in cavitation. The small amount of vapor generated is restricted to the nozzle inlet. Thus chamfering/rounding the orifice inlet geometry can inhibit cavitation by allowing a smoother entry to the orifice, and also improve the nozzle flow efficiency (C_d) as discussed below. This is due to the fact that flow uniformity in the orifice entrance region is significantly enhanced for the hydroground nozzle hence, cavitation is almost completely inhibited. This observation is consistent with those reported by other researchers. A 2D cut-plane was constructed passing though the mid-plane. This view also highlights the fact that the hydroground nozzle cavitates significantly less compared to the base nozzle.

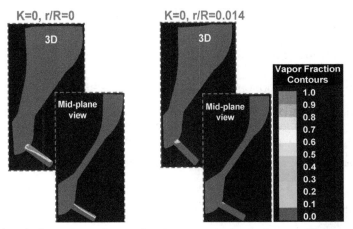

Fig. 4. 3D and mid-plane views of vapor fraction contours for the base and hydroground nozzles. Simulations were performed at P_{in}=1300bar, P_b=30bar, and full needle open position.

Figure 5 presents the velocity vectors plotted at the orifice inlet of the mid-plane for the base and hydroground nozzles presented in the context of previous figure. The zoomed view clearly shows that the velocity vectors point away from the wall for the base nozzle, while they are aligned with the flow for the hydroground nozzle thus ensuring a smooth entry into the orifice which decreases cavitation. As mentioned earlier, difference in cavitation characteristics plays a central role in spray breakup processes. Hence, spray behavior of a hydroground nozzle is expected to be different from that of the base nozzle.

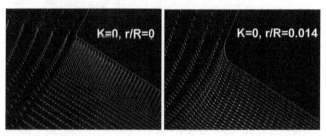

Fig. 5. Velocity vectors shown at the orifice inlet of the mid-plane for base and cylindrical nozzles presented in the context of Fig. 4.

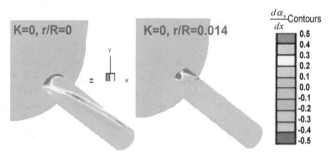

Fig. 6. Contours of $\frac{d\alpha_v}{dx}$ for the two injector orifices described in the context of Fig. 4

Figure 6 presents the contours of $\frac{d\alpha_v}{dx}$ where α_v is the fuel vapor fraction and 'x' is the coordinate axis along the orifice. Hence the parameter $\frac{d\alpha_v}{dx}$ represents the production or consumption of fuel vapor inside the orifice. Positive represents production while negative values indicate consumption of fuel vapor. A value of 0 represents no change in the fuel vapor fraction with axial position. A zoomed 3D view of the sac and upper orifice region is shown. For both the nozzles, the sac region is composed of pure liquid hence $\frac{d\alpha_v}{dx} = 0$. Since the vapor generation takes place at the upper side of orifice inlet it is not surprising that $\frac{d\alpha_v}{dx}$ is positive. The vapor fraction contours for the base nozzle showed pure vapor existence throughout the upper part of orifice (cf. Fig. 4). However, $\frac{d\alpha_v}{dx}$ only predicts pockets of vapor formation indicating that the remaining vapor is due to advection from the orifice inlet. In the case of conical (not shown here) and hydroground nozzles, vapor is generated at the orifice inlet however it is completely consumed soon after; hence the exit of the orifice is composed of pure liquid fuel only.

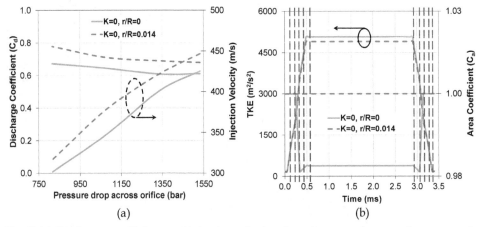

(a)   (b)

Fig. 7. (a) Discharge coefficient and injection velocity plotted versus pressure drop across the orifice, (b) Turbulence parameters such as TKE and TDR as a function of time for the base and hydrogroud nozzles shown in Figure 4.

Figure 7a presents C_d and injection velocity at nozzle exit for different pressure drops across the orifice. The back pressure was always fixed at 30bar, hence, the change in injection pressure resulted in change in pressure drop across the orifice. The methodology for calculating these parameters was discussed earlier. With increase in injection pressure, injection velocity at the orifice exit is seen to increase, which is expected. It should be noted that the injection velocity reported is an average value across the orifice. As expected, the average injection velocity and discharge coefficient is lower for the base nozzle owing to the presence of cavitation at the orifice exit. The influence of nozzle geometry on turbulence levels at the nozzle orifice exit is investigated in Fig. 7b since these parameters are directly input in spray simulations as rate profiles for cavitation, turbulence, and fuel mass injected. The different needle lift positions simulated are also shown. The peak needle lift of this injector was 0.275 mm which corresponds to full needle open position. Other needle positions simulated are: 0.05mm, 0.1mm, 0.15mm, and 0.2mm open respectively. A general trend observed is that the turbulent kinetic energy (TKE) increased with needle lift position which is expected since the injection pressure also increased resulting in higher Reynolds numbers. TKE and turbulent dissipation rate (TDR) were seen to be higher for the base nozzle case at all needle lift positions. Turbulence is known to play a key role in spray breakup processes; hence, accounting for such differences in turbulence levels between orifices is expected to improve spray predictions. The reason for similar turbulence levels at lower needle lifts is due to the fact that at low needle lift positions, the area between the needle and orifice governs the fluid dynamics inside the nozzle. However, at full needle lift position during the quasi-steady injection period, the orifice plays a critical role in the flow development inside the nozzle. Area coefficient was unity for the hydroground nozzle which is expected since this orifice inhibits cavitation inception completely. These rate profiles are input for the spray simulations (Som, 2009a, 2010b, 2011).

3.3 Influence of fuel properties on nozzle flow

This section presents the influence of fuel properties on nozzle flow development. As mentioned earlier, the nozzle flow characteristics of biodiesel is compared against that of diesel fuel since biodiesel is a lucrative blending agent. Table 2 presents the physical properties of diesel and biodiesel (soy-methyl ester) fuels. There are small differences in density and surface tension between these fuels. However, major differences are observed in viscosity and vapor pressure values. These differences are expected to influence the nozzle flow and spray development.

Fuel Property	Diesel	Biodiesel
Carbon Content [wt %]	87	76.74
Hydrogen Content [wt %]	13	12.01
Oxygen Content [wt %]	0	11.25
Density @ 15°C (kg/m^3)	822.7	877.2
Dynamic Viscosity @ 40°C (cP)	1.69	5.626
Surface Tension @ 25°C (N/m)	0.0020	0.00296
Vapor Pressure @ 25°C (Pa)	1000	1

Table 2. Comparison of physical properties of diesel and biodiesel (soy-methyl ester) fuels

Figure 8 presents the vapor fraction contours for diesel and biodiesel for P_{inj}=1300 bar and P_{back}=30 bar. The 3-D view of the cavitation contours indicates that vapor generation occurs at the orifice inlet for both the fuels. For diesel, these cavitation contours, generated at the upper side of the orifice, reach the orifice exit. In contrast, for biodiesel, the cavitation contours only extend a few microns into the orifice and do not reach the injector exit. Since cavitation plays a significant role in primary breakup, the atomization and spray behavior of these fuels is expected to be different. The mid-plane view also indicates that the amount of cavitation is significantly reduced for biodiesel compared to diesel. This is mainly due to two reasons:

1. The vapor pressure of biodiesel is lower than diesel fuel. Cavitation occurs when the local pressure is lower than the vapor pressure of the fuel. Hence, reduction in vapor formation can be expected for fuels with lower vapor pressures. Although injection pressures are very high, the differences in vapor pressure values are also important for cavitation inception.
2. The viscosity of biodiesel is higher compared to diesel fuel (cf. Table 2). This increased viscosity results in lower velocities inside the sac and orifice, which in turn decreases the velocity gradients. This also results in lowering of cavitation patterns for biodiesel.

Figure 9 presents contours of the magnitude of velocity at the mid-plane and orifice exit plane for diesel and biodiesel fuels for the case presented in Fig. 8. The flow entering the orifice encounters a sharp bend (i.e., large velocity and pressure gradients) at the upper side of the orifice inlet, causing cavitation in this region, as indicated by the vapor fraction contours. Upstream of the orifice, the velocity distribution appears to be similar for the two fuels. However, at the orifice exit, the contours indicate regions of higher velocity for diesel compared to biodiesel. This is related to the fact that the viscosity (cf. Table 2) of biodiesel is higher than that of diesel fuel. The velocity contours at the orifice exit indicate fairly symmetrical distribution with respect to the y-axis for both fuels.

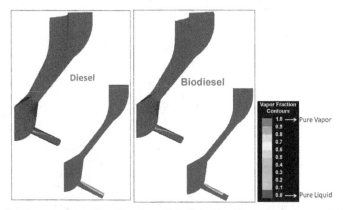

Fig. 8. Vapor fraction contours for diesel and biodiesel inside the injector and at the mid-plane. The simulations were performed at full needle open position with P_{inj} = 1300 bar and P_{back} = 30 bar.

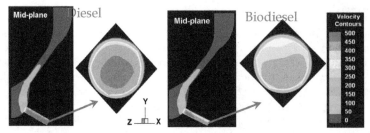

Fig. 9. Velocity contours for diesel and biodiesel at the mid-plane and orifice exit. The simulations were performed at full needle open position with P_{inj} = 1300 bar and P_{back} = 30 bar.

Figure 10 presents the computed fuel injection velocity, mass flow rate, C_d, and normalized TKE at the nozzle exit for different injection pressures. All these parameters are obtained by computing the 3-D flow inside the injector and then averaging the properties at the orifice exit. As expected, with increased injection pressure, the injection velocity and mass flow rate at the orifice exit increase (cf. Fig. 10a). However, the injection velocity, mass flow rate, and discharge coefficient are lower for biodiesel compared to diesel fuel. This difference in injection velocity and hence in mass flow rate can be attributed to the significantly higher viscosity of biodiesel. The lower mass flow rate for biodiesel implies that, for a fixed injection duration, a lesser amount of biodiesel will be injected into the combustion chamber compared to diesel. Combined with the lower heating value of biodiesel, this would lead to lower engine output with biodiesel compared to diesel fuel. As indicated in Fig. 10b, the average TKE at the nozzle exit is also lower for biodiesel. This is due to the fact that the Reynolds number is lower for biodiesel due to its higher effective viscosity. This has implications for the atomization and spray characteristics of the two fuels, since the turbulence level at the orifice exit influences the primary breakup.

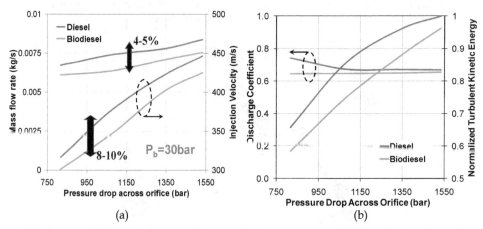

Fig. 10. Computed flow properties at the nozzle exit versus pressure drop in the injector for diesel and biodiesel fuels: (a) mass flow rate and injection velocity; (b) discharge coefficient and normalized TKE.

4. Conclusion

The flow inside the nozzle is critical in spray, combustion, and emission processes for an internal combustion engine. Inner nozzle flows are multi-scale and multi-phase in nature, hence, challenging to capture both in experiments and simulations. Cavitation and turbulence generated inside the nozzle is known to influence the primary breakup of the fuel, especially in the near nozzle region. The authors capture the in-nozzle flow development using the two-phase flow model in FLUENT software. The influence of definition of cavitation inception is first analyzed by implementing an improved criterion for cavitation inception under turbulent conditions. While noticeable differences between the standard and advanced criteria for cavitation inception are observed under two-dimensional flow conditions, thorough development and validation is necessary before implementing in real injection flow simulations.

Since the injector nozzle is a critical component of modern internal combustion engines, the influence of orifice geometry and fuel properties on in-nozzle flow development were also characterized. Both cavitation and turbulence was reduced using a hydroground nozzle compared to a base production nozzle. This will result in significant differences in spray, combustion, and emission behaviour also for these nozzles. Biodiesel being a lucrative blending agent for compression ignition engine applications was then compared to diesel fuel for inner nozzle flow development. Cavitation and turbulence generated inside the nozzle was observed to be lower for biodiesel compared to diesel fuel. Additionally, boundary conditions in terms of cavitation, turbulence, and flow variables were obtained for spray combustion simulations as a function of time for the detailed nozzle flow simulations.

5. Acknowledgment

The submitted manuscript has been created by UChicago Argonne, LLC, Operator of Argonne National Laboratory ("Argonne"). Argonne, a U.S. Department of Energy Office of Science laboratory, is operated under Contract No. DE-AC02-06CH11357. The U.S. Government retains for itself, and others acting on its behalf, a paid-up nonexclusive, irrevocable worldwide license in said article to reproduce, prepare derivative works, distribute copies to the public, and perform publicly and display publicly, by or on behalf of the Government.

6. References

Bae C., Yu J., Kang J., Kong J., Lee K.O. 2002. Effect of nozzle geometry on the common-rail diesel spray. *SAE Paper No.* 2002-01-1625.

Benajes, J., Pastor, J.V., Payri, R., Plazas, A.H. 2004. Analysis of the influence of diesel nozzle geometry in the injection rate characteristics. *Journal of Fluids Engineering*, 126, 63-71.

Blessing, M., Konig, G., Kruger, C., Michels, U., Schwarz, V. 2003. Analysis of flow and cavitation phenomena in diesel injection nozzles and its effects on spray and mixture formation. *SAE Paper No.* 2003-01-1358.

Bosch, W. 1966. The Fuel Rate Indicator: A New Measuring Instrument for Display of the Characteristics of Individual Injection. *SAE Paper No.* 660749.

Brennen, E.C. 1995. Cavitation and Bubble Dynamics. Oxford University Press.

Dabiri, S., Sirignano, W.A., Joseph, D.D. 2007. Cavitation in an Orifice Flow. *Physics of Fluids*, 19, pp. 072112-1-072112-9.

Han, J.S., Lu, P.H., Xie, X.B., Lai, M.C., Henein, N.A. 2002. Investigation of diesel spray primary breakup and development for different nozzle geometries. *SAE Paper No.* 2002-01-2775.

Hountalas, D.T., Zannis, T.C., Mavropoulos, G.C., Schwarz, V., Benajes, J., Gonzalez C.A. 2005. Use of a Multi-Zone Combustion Model to Interpret the effet of injector nozzle hole geometry on HD DI Diesel Engine performance and pollutant emissions. *SAE Paper No.* 2005-01-0367.

Joseph, D.D. 1998. Cavitation and the state of stress in a flowing liquid. *Journal of Fluid Mechanics*, 366, 367-378.

Kottke, P.A., Bair, S.S., Winer, W.O. 2005. Cavitation in creeping shear flows. *AICHE Journal*, 51, 2150.

Mulemane, A., Han, J.S., Lu, P.H., Yoon, S.J., Lai, M.C. 2004. Modeling dynamic behavior of diesel fuel injection systems. *SAE Paper No.* 2004-01-0536.

Naber, J.D., Siebers, D.L. 1996. Effects of Gas Density and Vaporization on Penetration and Dispersion of Diesel Sprays. *SAE Paper No.* 960034.

Padrino, J.C., Joseph, D.D., Funada, T., Wang, J., and Sirignano, W.A. 2007. Stress-Induced cavitation for the streaming motion of a viscous liquid past a sphere. *Journal of Fluid Mechanics*, 578, 381-411.

Payri, F., Bermudez, V., Payri, R., Salvador, F.J. 2004. The influence of cavitation on the internal flow and the spray characteristics in diesel injection nozzles. *Fuel*, 83, 419-431.

Payri, R., Garcia, J.M., Salvador, F.J., Gimeno, J. 2005. Using spray momentum flux measurements to understand the influence of diesel nozzle geometry on spray characteristics. *Fuel*, 84, 551-561.

Payri, R., Salvador, F.J., Gimeno, J., Zapata, L.D. 2008. Diesel nozzle geometry influence on spray liquid-phase fuel penetration in evaporative conditions. *Fuel*, 87, 1165-1176.

Payri, F., Margot, X., Patouna, S., Ravet, F., Funk, M. 2009. A CFD study of the effect of the needle movement on the cavitation pattern of diesel injectors. *SAE Paper No.* 2009-24-0025.

Singhal, A.K., Athavale, A.K., Li, H., Jiang, Y. 2002. Mathematical basis and validation of the full cavitation model. *Journal of Fluid Engineering*, 124, 617-624.

Som, S. Development and validation of spray models for investigating Diesel engine combustion and emissions. *PhD thesis*, University of Illinois at Chicago; 2009.

Som, S., Ramirez, A.I., Aggarwal, S.K., Kastengren, A.L., El-Hannouny, E.M., Longman, D.E., Powell, C.F., Senecal, P.K. 2009. Development and Validation of a Primary Breakup Model for Diesel Engine Applications. *SAE Paper No.* 2009-01-0838.

Som, S., Aggarwal, S.K. 2009. An Assessment of Atomization Models for Diesel Engine Simulations. *Atomization and Sprays*, 19(9), 885-903.

Som, S., Aggarwal, S.K., El-Hannouny, E.M., Longman, D.E. 2010. Investigation of nozzle flow and cavitation characteristics in a diesel injector. *Journal of Engineering for Gas Turbine and Power*, 132, 1-12.

Som, S., Longman, D.E., Ramírez, A.I., Aggarwal, S.K. 2010. A comparison of injector flow and spray characteristics of biodiesel with petrodiesel. *Fuel*, 89, 4014-4024.

Som, S., Aggarwal, S.K. 2010. Effects of primary breakup modeling on spray and combustion characteristics of compression ignition engines. *Combustion and Flame*, 157, 1179-1193.

Som, S., Ramirez, A.I., Longman, D.E., Aggarwal, S.K. 2011. Effect of nozzle orifice geometry on spray, combustion, and emission characteristics under diesel engine conditions. *Fuel*, 90, 1267-1276.

Winer, W.O., Bair, S. 1987. The Influence of Ambient Pressure on the Apparent Shear Thinning of Liquid Lubricants - An Overlooked Phenomenon. *Proc. Inst. Mech. Eng.*, C190-87, pp. 395-398.

Winklhofer, E., Kull, E., Kelz, E., and Morozov, A. 2001. Comprehensive Hydraulic and flow field documentation in model throttle experiments under cavitation conditions. *ILASS Europe*, 2001.

6

Numerical Simulation of Biofuels Injection

Jorge Barata and André Silva

Aerospace Sciences Department, University Beira Interior,
Covilhã,
Portugal

1. Introduction

The use of alternative biofuels in the co-generation of electricity and heat, as well as in the transportation sector, presents major benefits, such as the conservation of the environment due to their renewable origin, the reduction of fossil fuels use or the conservation of agricultural activity in regions where the food production is being reduced.

The more important biofuels currently under investigation are the bio-alchohols and their derived ethers, and the vegetable oils and their derived esters. Methyl esters of rapeseed oils or soybean oils have been tested in Diesel engines, and in spite of the strong dispersion of the published results, there are indications that their use is a promising solution to the problems originated with the raw vegetable oil due to their higher viscosity, boiling temperature, final temperature of distillation and point of obstruction of cold filter (Tinaut, 2005).

The present work presents a numerical study on evaporating biofuel droplets injected through a turbulent cross-stream. This study uses an Eulerian/Lagragian approach to account for turbulent transport, dispersion, evaporation and coupling between both processes in practical spray injection systems, which usually include air flows in the combustion chamber like swirl, tumble and squish in I.C. engines or crossflow in boilers and gas turbines. An array of evaporating biofuel droplets through a crossflow is studied, and a comparison of the droplet fuel dispersion and evaporation with conventional fuels is performed. A summary of the main general characteristics properties of the conventional fuels and biofuels tested in the present investigation is presented in Table 1.

The evaporation of droplets in a spray involves simultaneous heat and mass transfer processes in which the heat for evaporation is transferred to the drop surface by conduction and convection from the surrounding hot gas, and vapour is transferred by convection and diffusion back into the gas stream. The overall rate of evaporation depends on the pressure, temperature, and transport properties of the gas; the temperature, volatility and diameter of the drops in the spray; and the velocity of the drops relative to that of the surrounding gas (Faeth, 1983, 1989, 1986).

Godsave (1953) and Spalding (1953) gave the basic droplet vaporization/combustion model for an isolated single-component droplet in a stagnant environment. Since then this model has been studied extensively both experimentally and theoretically. These studies have been reviewed extensively by several authors during the past decades (e.g. Williams, 1973; Faeth, 1977; Lefebvre 1989; Law, 1982; and Sirignano, 1978), and are mostly dedicated to study the dynamics of a single droplet. Abramzon and Sirignano (1989) and Berlemont et al. (1995) presented droplet vaporization models for the case of a spray in stagnant surroundings, and showed that the convective effects were most relevant. The same type of configuration was studied by Chen and Pereira (1992), and the predictions were found to follow satisfactorily the measurements presented. More recently, Sommerfeld (1998) presented a study on stationary turbulent sprays, using a droplet evaporation model based on the model proposed by Abramzon and Sirignano (1989), and revealed a general good agreement with experiments.

	n-Heptane	Rapeseed Methyl Ester (RME)	Diesel Fuel (DF-2)	Ethanol
Molecular mass, M_F	100.16	300	198	46
Fuel density at 288.6°K, $\rho_{F288.6K}$	687.8 Kg/m³	880 Kg/m³	846 Kg/m³	790 Kg/m³
Boiling temperature at atmospheric pressure, T_{bn}	371.4°K	613°K	536.4°K	351°K
Latent heat of vaporization at atmospheric pressure, L_{Tbn}	371,8KJ/Kg	254KJ/Kg	254KJ/Kg	904KJ/Kg

Table 1. Characteristic properties of biofuels compared with n-Heptane and Diesel Fuel.

If special attention is dedicated to the biofuels injection and evaporation, then practically no numerical or experimental studies can be found. Recently Bai et al. (2002) presented a most relevant numerical study of a spray in wind tunnel using the Arcoumanis et al. (1997) experiments, but concentrated on the development of the spray impingement model and the fuel used was gasoline.

2. Mathematical model

This section describes the mathematical model for turbulent particle dispersion and vaporization assuming that the particles are sufficiently dispersed so that particle-particle interaction is negligible.

The particle phase is described using a Lagrangian approach while an Eulerian frame is used to describe the effects of both interphase slip and turbulence on particle motion using random-sampling techniques (Monte Carlo). It is also assumed that the mean flow is steady and the material properties of the phases are constant.

When vaporizing droplets are involved in the simulations, two-way coupling must be accounted for since the phase change modifies the characteristics of the fluid phase. The vapour produced by the droplets is a mass source for the fluid; moreover the vaporization process generates modifications in the momentum and energy balances between both phases. Fluid phase equations then contain many extra-source terms. It is assumed that the vapour production does not significantly modify the fluid phase density.

The method to solve the continuous phase is based on the solution of the conservation equations for momentum and mass. Turbulence is modelled with the "k-ε" turbulence model of Launder and Spalding (1974), which is widely and thoroughly tested, and was found to predict reasonably well the mean flow (Barata, 1998). In order to reduce the numerical errors to an acceptable level, the higher-order QUICK scheme of Leonard (1979) is used to evaluate the convection terms. A similar method has been used for three-dimensional (Barata, 1998) or axisymmetric flows (Shuen et al., 1985; Lilley, 1976; Lockwood & Naguib, 1975) and only the main features are summarized here.

The governing equations (continuity, momentum, turbulent kinetic energy, dissipation, enthalpy, and vapour mass fraction) constitute a set of coupled partial differential equations that can be reduced to a single convective-diffusive conservation equation of the form

$$\frac{\partial(\rho U_i \phi)}{\partial X_i} = \frac{\partial}{\partial X_i}\left[\Gamma_\varphi \frac{\partial \phi}{\partial X_i}\right] + S_\varphi \tag{1}$$

where Γ_ϕ is the effective diffusion coefficient for quantity ϕ. The term on the left-hand side is the convection term, whilst the first and the second terms on the right-hand side are the diffusion term and the source term, respectively.

The source term S_ϕ as divided into two parts, which yields the following expression:

$$S_\phi = S_{\phi g} + S_{\phi p} \tag{2}$$

where $S_{\phi g}$, specifies the source term of the gas and $S_{\phi p}$, specifies the source term of the particle.

ϕ	$S_{\phi g}$	$S_{\phi p}$	Γ_ϕ
1	-	$S_{\rho,p}$	-
U_i	$-\frac{\partial}{\partial X_i}(P + \frac{2}{3}k) - \frac{\partial}{\partial X_j}\frac{2}{3}\mu_t \frac{\partial \overline{U_j}}{\partial X_i} + \rho g_i$	$S_{U_i,p}$	$\mu + \mu_T$
T	0	$S_{T,p}$	$\frac{\mu}{Pr} + \frac{\mu_T}{Pr_T}$
Y_p	0	$S_{Y_k,p}$	$\frac{\mu}{Sc} + \frac{\mu_T}{Sc_T}$
k	$G - \rho\varepsilon$	$S_{k,p}$	$\mu + \frac{\mu_T}{\sigma_k}$
ε	$C_{\varepsilon 1}\frac{\varepsilon}{k}G - C_{\varepsilon 2}\rho\varepsilon$	$S_{\varepsilon,p}$	$\mu + \frac{\mu_T}{\sigma_\varepsilon}$

Table 2. Terms in the general form of the differential equation.

The source terms of the gas phase, $S_{\phi g}$ and the effective diffusion coefficient Γ_ϕ, are summarized in table 2 for different depended variables. G is the usual turbulence energy production term defined as:

$$G = \mu_t \left[\frac{\partial \overline{U_i}}{\partial X_j} + \frac{\partial \overline{U_j}}{\partial X_i} \right] \frac{\partial \overline{U_i}}{\partial X_j} \tag{3}$$

and

$$\mu_t = C_\mu \rho \frac{k^2}{\varepsilon} \tag{4}$$

The turbulence model constants that are used are those indicated by Launder and Spalding (1974) that have given good results for a large number of flows, and are summarized in the next table.

$C\mu$	$C_{\varepsilon 1}$	$C_{\varepsilon 2}$	σ_k	σ_ε	$C_{\varepsilon 3}$	Pr_t	Sc_t	Pr	Sc
0.09	1.44	1.92	1.0	1.3	1.1	0.6	0.85	$\mu C_P / k_g$	$\mu / \rho D$

Table 3. Turbulence model constants.

Vaporization phenomena are described in the present study assuming spherical symmetry for heat and mass transfers between the droplet and the surrounding fluid, and convection effects are taken into account by introducing empirical correlation laws.

The main assumptions of the models are: spherical symmetry; quasi-steady gas film around the droplet; uniform physical properties of the surrounding fluid; uniform pressure around the droplet; and liquid/vapor thermal equilibrium on the droplet surface.

The effect of the convective transport caused by the droplet motion relative to the gas was a accounted for by the so called "film theory", which results in modified correlations for the Nusselt and Sherwood numbers. For rapid evaporation (i.e boiling effects) additional corrections were applied. The infinite droplet conductivity model was used to describe the liquid side heat transfer taking into account droplet heat-up. Hence, the following differential equations for the temporal changes of droplet size and temperature have to be solved.

$$\frac{dD_p}{dt} = -\frac{2\dot{m}}{\pi \rho_F D_p^2} \tag{5}$$

$$\frac{dT_p}{dt} = \frac{6Q_L}{\pi \rho_F C_{P_F} D_p^3} \tag{6}$$

Under the assumption of steady state conditions in the gas film and assuming a spherical control surface around the droplet, the total mass flow through this surface will be equal to the evaporation rate $\dot{m}$:

$$\dot{m} = \pi \overline{\rho_g D_g} D_p Sh * \ln(1 + B_M)$$ (7)

and

$$\dot{m} = \pi \frac{\overline{K_{vap}}}{C_{pvap}} D_p Nu * \ln(1 + B_T)$$ (8)

the quantity $\overline{\rho_g D_g}$ can be replaced with K_{vap}/C_{pvap}, assuming a Lewis number of unity.

The heat penetrating into the droplet can be expressed by:

$$Q_L = \dot{m}\left(\frac{\overline{C_{vap}}(T_\infty - T_s)}{B_M} - L(T_s)\right)$$ (9)

The mass transfer number B_M as defined as

$$B_M = \frac{Y_{Fs} - Y_{F\infty}}{1 - Y_{Fs}}$$ (10)

where Y_{Fs} is the fuel mass fraction on the droplet surface and defined as:

$$Y_{Fs} = \left[1 + \left(\frac{P}{P_{Fs}} - 1\right)\frac{M_A}{M_F}\right]^{-1}$$ (11)

For any given value of surface temperature, the vapor pressure is readily estimated from the Clausius-Claperyon equation as

$$P_{Fs} = \exp\left(a - \frac{b}{T_s - 43}\right)$$ (12)

where a and b are constants of the fuel.

The latent heat of vaporization is given by Watson (1931) as

$$L(T_s) = L_{tbn}\left(\frac{T_{cr} - T_s}{T_{cr} - T_{bn}}\right)^{-0.38}$$ (13)

Equations 7 and 8 for $\dot{m}$ are similar to the expressions for the droplet vaporization rate predicted by the classical model, with the values of the non-dimensional parameters Nu_0 and Sh_0 in the classical formulas substituted by $Nu*$ and $Sh*$ respectively. Where are expressed as

$$Sh* = 2 + (Sh_0 - 2)/F_M$$ (14)

$$Nu* = 2 + (Nu_0 - 2)/F_T$$ (15)

The parameters Nu^* and Sh^* are the "modified" Nusselt and Sherwood numbers, and tend to Nu_0 and Sh_0, respectively, as F_T and F_M tend to the unity.

In the case of an isothermal surface and constant physical properties of the fluid, the problem has a self-similar solution and the correction factors F_M and F_T do not depend on the local Reynolds number. It was found that the values F_M and F_T are practically insensitive to the Schmidt and Prandtl numbers and the wedge angle variations, and can be approximated as

$$F_M - \Gamma(B_M), \quad \Gamma_T - \Gamma(B_l) \tag{16}$$

where $F(B)$ is given by

$$F(B) = (1+B)^{0.7} \frac{\ln(1+B)}{B} \tag{17}$$

Nu_0 and Sh_0 are evaluated by the Frossling correlations:

$$Nu_0 = 2 + 0.552 \, Re^{1/2} \, Pr^{1/3} \tag{18}$$

$$Sh_0 = 2 + 0.552 \, Re^{1/2} \, Sc^{1/3} \tag{19}$$

The evaporation rate $\dot{m}$ with convection is:

$$\dot{m} = \pi \overline{\rho_g D_g} D_p \ln(1+B_M) \left(2 + \frac{0.552 \, Re^{1/2} \, Sc^{1/3}}{F_M} \right) \tag{20}$$

and

$$\dot{m} = \pi \frac{\overline{K_{vap}}}{C_{pvap}} D_p \ln(1+B_T) \left(2 + \frac{0.552 \, Re^{1/2} \, Pr^{1/3}}{F_T} \right) \tag{21}$$

The Schmidt number and the Prandtl number are equal assuming a Lewis number of unity. Equation 20 has the advantage that it applies under all conditions, including the transient state of droplet heat-up, whereas Eq. (31) can only be used for steady-state evaporation.

Finally the evaporation rate $\dot{m}$ is:

$$\dot{m} = 2\pi \frac{\overline{K_{vap}}}{C_{pvap}} D_p \ln(1+B_M) \left(1 + \frac{0.276 \, Re^{1/2} \, Pr^{1/3}}{F_M} \right) \tag{22}$$

And the equations for the temporal changes of droplet size and temperature are:

$$\frac{dD_p}{dt} = -\frac{4K_{vap}\ln(1+B_M)}{C_{pvap}\rho_F D_p}\left(1+\frac{0.276\,\mathrm{Re}^{\frac{1}{2}}\,\mathrm{Pr}^{\frac{1}{3}}}{F_M}\right) \tag{23}$$

$$\frac{dT_p}{dt} = \frac{12K_g\ln(1+B_M)}{Cv_{ap}\rho_F D_p^2 C_{P_F}}\left(1+\frac{0.276\,\mathrm{Re}^{\frac{1}{2}}\,\mathrm{Pr}^{\frac{1}{3}}}{F_M}\right)\left(\frac{C_{vap}(T_\infty-T_s)}{B_M}-L(T_s)\right) \tag{24}$$

Of the air/vapor mixture in the boundary layer near the droplet surface according to Hubbard et al. (1973), the best results are obtained using the one-third role (Sparrow & Gregg, 1958), where average properties are evaluated at the following reference temperature and composition:

$$T_r = T_s + \frac{T_\infty - T_s}{3} \tag{25}$$

$$Y_{Fr} = Y_{Fs} + \frac{Y_{F\infty} - Y_{Fs}}{3} \tag{26}$$

For example, the reference specific heat at constant pressure is obtained as

$$C_{p_{vap}} = Y_{Ar}\left(C_{p_{Ar}} \quad at \quad T_r\right) + Y_{Fr}\left(C_{p_F} \quad at \quad T_r\right) \tag{27}$$

The dispersed phase was treated using the Lagrangian reference frame. Particle trajectories were obtained by solving the particle momentum equation through the Eulerian fluid velocity field, for a sufficiently high number of trajectories to provide a representative statistics.

The equations used to calculate the position and velocity of each particle were obtained considering the usual simplification for dilute particle-laden flows (Shuen et al., 1985). Static pressure gradients are small, particles can be assumed spherical and particle collisions can be neglected. Since $\rho_p/\rho_f\rangle 200$, the effects of Basset, virtual mass, Magnus, Saffman and buoyancy forces are negligible (Arcoumanis et al., 1997; Lockwood & Naguib, 1975). In dilute flows of engineering interest, the steady-state drag term is the most important force acting on the particle. Under these conditions the simplified particle momentum equation is:

$$\frac{\partial u_{p;i}}{\partial t} = \frac{1}{\tau_p}\left(u_{f;i}-u_{p;i}\right)+g_i \tag{28}$$

The mathematical expression for the relaxation time, τ_p, is

$$\tau_p = \frac{24\rho_p D_p^2}{18\mu_f C_D\,\mathrm{Re}_p} \tag{29}$$

where Re_p is the particle Reynolds number,

$$\mathrm{Re}_p = \frac{\rho_f\left|\overrightarrow{V_p}-\overrightarrow{V_f}\right|D_p}{\mu_f} \tag{30}$$

Note that the physical properties of ρ_f and μ_f should be evaluated at the reference temperature T_r and are

$$\mu_f = Y_{Ar}\left(\mu_A \text{ at } T_r\right) + Y_{Fr}\left(\mu_F \text{ at } T_r\right) \tag{31}$$

$$\rho_{vap} = \left(\frac{Y_{Ar}}{\rho_A} + \frac{Y_{Fr}}{\rho_F}\right)^{-1} \tag{32}$$

and C_D is the drag coefficient (Shirolkar et al., 1996),

$$C_D = \left(\frac{24}{\text{Re}_p}\right)\left(1 + 0.15\,\text{Re}_p^{0.687}\right) \tag{33}$$

for $\text{Re}_p < 10^3$.

The particle momentum equation can be analytically solved over small time steps, Δt, and the particle trajectory is given by

$$u_{p;i}^{NEW} = u_{f;i} + \left(u_{p;i}^{OLD} - u_{f;i}\right)e^{-\Delta t/\tau_p} + g_i\tau_p\left[1 - e^{-\Delta t/\tau_p}\right] \tag{34}$$

$$x_{p;i}^{NEW} = x_{p;i}^{OLD} + \frac{\Delta t}{2}\left(u_{p;i}^{NEW} + u_{p;i}^{OLD}\right) \tag{35}$$

The critical issues are to determine the instantaneous fluid velocity and the evaluation of the time, Δt, of interaction of a particle with a particular eddy.

The time step is obviously the eddy-particle interaction time, which is the minimum of the eddy lifetime, τ_{FL}, and the eddy transit time, t_c. The eddy lifetime is estimated assuming that the characteristic size of an eddy is the dissipation length scale in isotropic flow:

$$l_e = B\frac{k^{3/2}}{\varepsilon} \approx C_\mu^{3/4}\frac{k^{3/2}}{\varepsilon} \tag{36}$$

$$\tau_{FL} = A\frac{k}{\varepsilon} \approx 0.2\frac{k}{\varepsilon} \tag{37}$$

where A and B are two dependent constants (Shirolkar et al., 1996).

The transit time, t_c, is the minimum time a particle would take to cross an eddy with characteristic dimension, l_e, and is given by

$$t_c = \frac{l_e}{\left|\overrightarrow{v_d}\right|} \tag{38}$$

where $\overrightarrow{v_d}$ is the relative velocity between the particle and the fluid (drift velocity).

A different expression for the transit time is also recommended in the literature (Shitolkar et al., 1996; Shuen et al., 1983; Gosman & Ioannides, 1981), and was used in the present work:

$$t_c = -\tau_p \ln\left(1 - \frac{l_e}{\tau_p \left|u_{f;i} - u_{p;i}\right|}\right) \tag{39}$$

where the drift velocity is also estimated at the beginning of a new iteration.

This equation has no solution when $l_e > \tau_p \left|u_{f;i} - u_{p;i}\right|$, that is, when the linearized stopping distance of the particle is smaller than the eddy size. In such a case, the particle can be assumed to be trapped by the eddy, and the interaction time will be the eddy lifetime.

The instantaneous velocity at the start of a particle-eddy interaction is obtained by random sampling from an isotropic Gaussian *pdf* having standard deviations of $\sqrt{2/3k}$ and zero mean values.

The above isotropic model was extended in the present work to account for cross-correlation's and anisotropy. To obtain the fluctuating velocities u'_f and v'_f, two fluctuating velocities u'_1 and u'_2 are sampled independently, and then are correlated using the correlation coefficient R_{uv}:

$$u'_f = u'_1 \tag{40}$$

$$v'_f = R_{uv} u'_1 + \sqrt{1 - R_{uv}^2}\, u'_2 \tag{41}$$

where $R_{uv} = \dfrac{\overline{u'_f v'_f}}{\sqrt{\overline{u'^2_f}}\sqrt{\overline{v'^2_f}}}$ was obtained from the measurements.

The interaction between the continuous and dispersed phase is introduced by treating particles as sources of mass, momentum and energy to the gaseous phase. The source terms due to the particles are calculated for each Eulerian cell of the continuous phase and are summarized in Table 4, and can be divided into two parts, which yields the following expression:

$$S_{\phi p} = S_{\phi i} + S_{\phi m} \tag{42}$$

where $S_{\phi i}$ specifies the source term due to inter-phase transport and $S_{\phi m}$ takes into consideration the transfer caused by evaporation.

To represent the temporal changes of droplet size and temperature Chen and Pereira (1992) used the following equations.

$$\frac{dD_p}{dt} = -\frac{4 K_{vap}}{C_{p_{vap}}} \ln\left(1 + \frac{C_{p_{vap}}}{L(T_s)}(T_\infty - T_s)\right)\left(1 + 0.23 \mathrm{Re}^{\frac{1}{2}}\right) \tag{43}$$

$$\frac{dT_p}{dt} = 12 K_g * \left(\frac{(T_\infty - T_s)}{\rho_F D_p^2 C_{P_F}}\right) * \left(1 + 0.3 \mathrm{Re}^{\frac{1}{2}} * \mathrm{Pr}^{\frac{1}{3}}\right) \tag{44}$$

In the last equation is assumed that the prevailing mode of heat transfer is forced convection, no evaporation occurs during the preheating period and the temperature is uniform across the droplet radius. For the forced convection the Ranz and Marshall (1952) correlation has taken the place of the Nusselt Number.

The solution of the governing equations was obtained using a finite-difference method that used discretized algebraic equations deduced from the exact differential equations that they represent. In order to reduce the numerical diffusion errors to an acceptable level, the quadratic upstream-weighted interpolation scheme was used (Leonard, 1979). Nevertheless, the usual grid independence tests were performed.

$S_{\phi p}$	$S_{\phi i}$	$S_{\phi m}$
$\overline{S_{\rho,p}}$	0	$\sum_p \dfrac{\dot{m}_p N_p}{V_{i,j}}$
$\overline{S_{U_i,p}}$	$-\sum_p \dfrac{\dot{m}_p N_p}{V_{i,j}} \left[\left(u_{j,p}^{t+\Delta t} - u_{j,p}^{t} \right) - g_i \Delta t \right]$	$\sum_p \dfrac{\dot{m}_p N_p u_{ia}}{V_{i,j}}$
$\overline{S_{T,p}}$	$-\sum_p \dfrac{N_p}{V_{i,j}} \left(\dfrac{L_{tbn} \dot{m}_p + Q_L}{C_{P_A}} \right)$	$\sum_p \dfrac{\dot{m}_p N_p}{V_{i,j}} \left(\dfrac{C_{P_{vap}}(T_s)*(T_r - T_s)}{C_{P_A}} \right)$
$\overline{S_{Y1,p}}$	0	0
$\overline{S_{Y2,p}}$	0	$\sum_p \dfrac{\dot{m}_p N_p}{V_{i,j}}$
$\overline{S_{k,p}}$	$\overline{U_j S_{Uji}} - \overline{U}_j \ \overline{S_{Uji}}$	$\overline{U_j S_{Ujm}} - \overline{U}_j \ \overline{S_{Ujm}} + \dfrac{1}{2}\overline{U}_j \ \overline{U}_j \ \overline{S_m} - \dfrac{1}{2}\overline{U_j U_j S_m}$
$\overline{S_{\varepsilon,p}}$	$C_{\varepsilon 3} \dfrac{\varepsilon}{k} \overline{S_{ki}}$	$C_{\varepsilon 3} \dfrac{\varepsilon}{k} \overline{S_{km}}$

Table 4. Dispersed phase source terms.

The computational domain (see Fig. 1) has six boundaries where dependent values are specified: an inlet plane and outlet planes, a symmetry plane, and three solid walls at the top, bottom and side of the channel. At the inlet boundary, uniform profiles of all dependent variables are set, while at the outflow boundaries, the gradients of dependent variables in the perpendicular direction are set to zero. On the symmetry plane, the normal velocity vanishes, and the normal derivates of the other variables are zero. At the solid surfaces, the wall function method described in detail by Launder and Spalding (1974) is used to prescribe the boundary conditions for the velocity and turbulence quantities, assuming that the turbulence is in state of local equilibrium.

The cross section of the computational domain is 0.05 x 0.05 m, whilst the channel length is 0.273 m. The droplets injection is perpendicular to the crossflow and the location of the injection point is 0.023 m far from the inlet plane ($Z_{in}/H = 0.46$).

The monosize array of droplets of 230μm of diameter is injected with an initial velocity V_p=-1m/s and a temperature of 293K or 443K through a crossflow with W_c=10m/s. The wall temperatures are 800K.

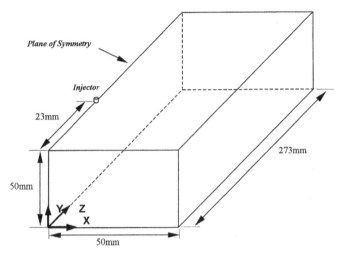

Fig. 1. Flow configuration

3. Results

To assess the computational method two sources of experimental data were used for the case of a polydisperse spray with a co-axial flow at atmospheric pressure (Heitor & Moreira, 1994) or elevated pressures (Barros, 1997). The method yielded reasonable results and revealed capabilities to improve the knowledge of the particle dispersion phenomena in more complex configurations. An example of the results obtained is shown in Figure 2, and a more complete analysis can be found in previous publications (Barata et al., 1999; Barata & Silva, 2000). The method was then extended to the case of an evaporating spray in a crossflow, and the evaporation models used by Chen and Pereira (1992) and Sommerfeld (1998) were tested and compared with the present model (see Barata, 2005 for details).

Figure 3 presents a parallel projection of the droplet trajectories in the vertical plane of symmetry (X=0) for two volatile fuels: n-Heptane and Ethanol. The former is used to define the zero limit of the anti-knock (resistance to pre-ignition) quality of fuels, while the other can be used to increase the octane number of gasoline. The higher volatility level of Ethanol can be inferred from Fig.3b) by the more uniform distribution at the right side of the domain and the trajectories in the direction of the top wall (at Y=0.05m). Due to the high volatility level of both fuels the droplets are injected and start almost immediately to evaporate, which gives rise to smaller droplets that follow quite closely the gaseous flow. Further downstream of the injection point, the trajectories of the droplets of Ethanol are more directed downwards than those of n-Heptane due to the higher fuel density and higher latent heat of vaporization. As a consequence, although a colder region near the injector is observed with Ethanol, the domain shows in general a much more uniform temperature distribution.

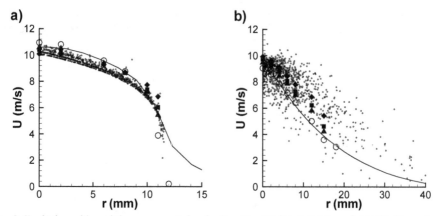

Fig. 2. Radial profiles of the mean axial velocity, U, at $X/D=0.2$ (a), and 6.5 (b). Experiments (Heitor & Moreira, 1994): ●, 30-35µm; ■, 40-45µm; ♦, 60-65µm; O, gaseous phase. Predictions: °, particles; ——, gaseous phase.

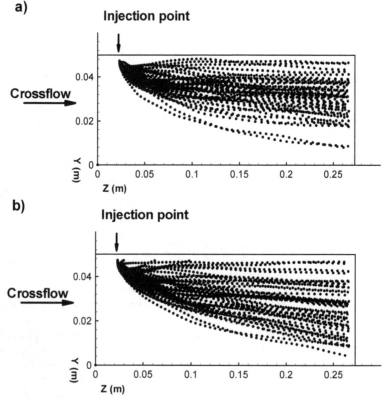

Fig. 3. Parallel projection of droplet trajectories in the vertical plane of symmetry ($X=0$) for $W_{cross}=10m/s$, $V_p=-1m/s$ and $T_p=293K$: (a) n-Heptane; (b) Ethanol.

Figure 4 shows the droplet temperature and diameter variation time for the different fuels and test conditions used in the present work that are summarized in Table 5. The horizontal part of the line of the temperature variation with time (Fig.4a) reveals the equilibrium of the evaporation process that corresponds to the horizontal part of the droplet diameter variation with time. It should be pointed out that since the droplet is moving in the direction of the Z coordinate (with the crossflow), the ambient temperature may not be constant, and the evolution of the droplet diameter with time is also influenced by its velocity.

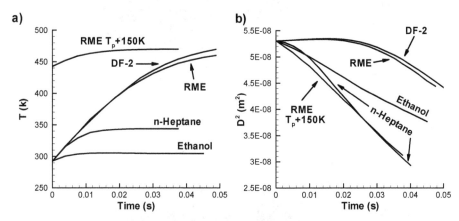

Fig. 4. Droplet temperature (a) and diameter (b) variation with time (W_{cross}=10m/s and V_p=-1m/s and T_p=293K).

The Rapeseed Methyl Ester (RME) and the Diesel Fuel (DF-2) have a higher boiling temperature and do not attain the equilibrium temperature in the first 50miliseconds (Fig.5a). As a consequence of the "pre-heating period" (about 20miliseconds in Fig.4b) the droplets diameters remain approximately constant and the evaporation starts later (Fig.4b).

Figure 5 shows parallel projections of the droplet trajectories in the vertical plane of symmetry, and confirms the main evaporation characteristics of the DF-2 and RME described in the previous paragraph. The pattern is similar for DF-2 and RME, although in the latter case there is a higher concentration of droplets in the core of the deflected monosize spray. This result is consistent with the slightly poorer evaporation characteristics of the RME deducted from Fig.5, and taking into account the average time that a droplet takes to reach the right hand side of the domain (at Z=0.3m) it is expected that its diameter would be (in average) at the exit of the channel about 92.2% of the initial diameter. So, in general the diameters of the droplets will be larger with DF-2 and RME, and the dispersion will be more difficult, because the slip between the gaseous phase and the dispersed phase will be more pronounced. Some collisions with the bottom wall are observed, but were not taken into account in the present study, although this phenomena has been investigated and reported elsewhere (see Barata & Silva, 2005).

Increasing the injection temperature of RME improves the evaporation of the droplets, and a more uniform distribution is obtained (Fig.5c). As shown in Fig.4, to obtain the equilibrium stage of evaporation near the injection point a pre-heating of 150K is required, which will be particularly difficult to implement in most of the practical situations.

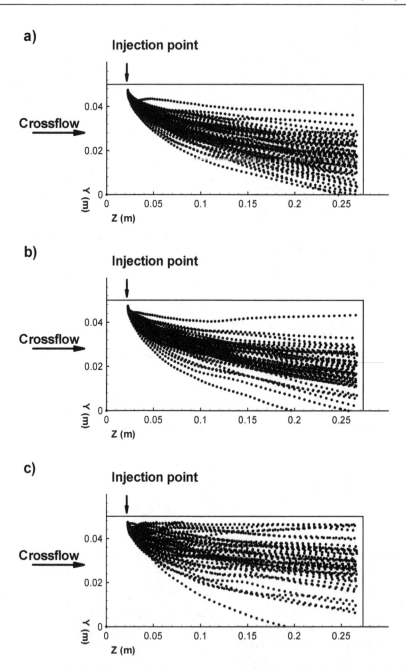

Fig. 5. Parallel projection of droplet trajectories in the vertical plane of symmetry (X=0) for W_{cross}=10m/s and V_p=-1m/s and T_p=293K.: (a) Diesel fuel (DF-2), T_p=293K; (b) RME, T_p=293K, (c) RME, T_p=443K.

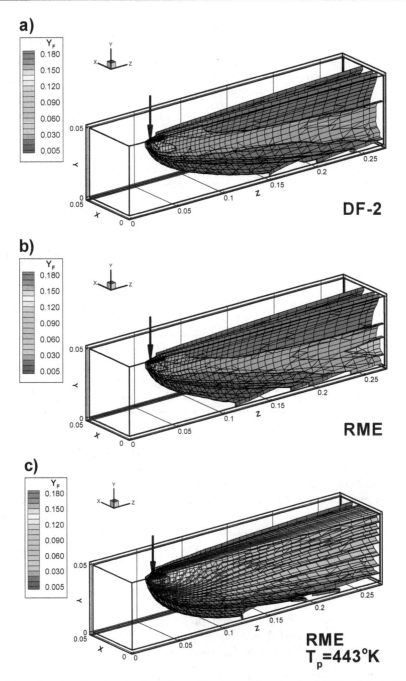

Fig. 6. Mass fraction distribution for W_{cross}=10m/s and V_p=-1m/s: (a) Diesel fuel (DF-2),
T_p=293°K; (b) RME, T_p=293°K, (c) RME, T_p=443°K.

Fuel	n-Heptane	Rapeseed Methyl ester (RME)	Rapeseed Methyl ester (RME)	Diesel fuel (DF-2)	Ethanol
Droplets diameter, d	230,3μm	230,3μm	230,3μm	230,3μm	230,3μm
Crossflow velocity, $W_{crossflow}$	10m/s	10m/s	10m/s	10m/s	10m/s
Crossflow temperature, $T_{crossflow}$	800°K	800°K	800°K	800°K	800°K
Temperature of the walls, T_{walls}	800°K	800°K	800°K	800°K	800°K
Droplet injection velocity, V_p	-1 m/s	-1 m/s	-1 m/s	-1 m/s	-1 m/s
Droplet injection temperature, T_p	293°K	293°K	443°K	293°K	293°K
Ambient pressure	1 bar 101 350Pa	1 bar 101 350Pa	1 bar 101 350Pa	1 bar 101 350Pa	1 bar 101 350Pa
Mass loading	20x10⁻⁴ Kg/s	20x10⁻⁴ Kg/s	20x10⁻⁴ Kg/s	20x10⁻⁴ Kg/s	20x10⁻⁴ Kg/s

Table 5. Summary of test conditions.

The evaporative characteristics of RME can be further analysed with the help of Fig.6 that shows a three-dimensional perspective of the mass fraction distribution with and without additional pre-heating. The results obtained with DF-2 and RME without additional pre-heating (Figs.6a and b) show that the mass fraction of fuel is always less than 0.04. For the DF-2 there is a larger evaporation near the injector, but further downstream the RME gives the higher values. When the additional pre-heating of 150K is used with RME, the domain shows a large region with a concentration of fuel vapour greater than 0.06, and the resulting pattern is quite similar to those obtained with more volatile fuels such as n-Heptane or Ethanol.

4. Conclusion

An Eulerian/Lagragian approach has been presented to calculate evaporating sprays through a crossflow. A method developed to study isothermal turbulent two- and three-dimensional dispersion was extended to the case of an array of evaporating biofuel droplets.

The droplet diameter, temperature and mass fraction distributions were found to be strongly dependent on the fuels used. Rapeseed Methyl Esters exhibit similar evaporating characteristics to DF-2, which indicates that it can be successfully used as an alternative fuel in many applications that utilize diffusion flames. The use of RME in homogeneous combustion systems may require a prohibitive level of pre-heating, and the use of Ethanol (obtained from sugar or starch crops) may be a better alternative.

5. Acknowledgements

The present work has been performed in the scope of the activities of the AFTUR project. The financial support of the European Commission under Contract n° ENK5-CT-2002-00662 is gratefully acknowledged.

6. References

Abramzon, B. and Sirignano, W.A., Droplet Vaporization Model for Spray Combustion Calculations, *Int. J. Heat Mass Transfer*, vol.12, no.9, 1989, pp.1605-1648.

Arcoumanis, C., Whitelaw, D.S. and Whitelaw, J.S., "Gasoline Injection against Surfaces and Films, Atomization and Sprays, vol.7, pp.437-456, 1997.

Bai, C.X., Rusche, H. and Gosman, "Modeling of Gasoline Spray Impingement", *Atomization and Sprays*, vol.12, pp.1-27, 2002.

Barata, J.M.M., Lopes, P.N.S.D. and Perestrelo, N.F.F., "Numerical Simulation of Polydisperse Two-Phase Turbulent Jets". AIAA Paper 99-3760, 30th AIAA Fluid Dynamics Conference, Norfolk, VA, 28 June-1 July, 1999.

Barata, J.M.M. and Silva, A.R.R., "Numerical Study of Spray Dispersion in a Premixing Chamber for Low-NOx Engines". Millennium International Symposium on Thermal and Fluid Sciences, Xi'an, China, 18-22 September, 2000.

Barata, J.M.M. and Silva, A.R.R., "The Impingement of a Deflected Spray". Eighth International Conference on Energy for a Clean Environment, Lisbon, Portugal, 27-30 June, 2005.

Barros, A., "Projecto e Construção de um Labotatório de Atomização a Alta Pressão", MSc Thesis, *Instituto Superior Técnico*, Lisbon, December, 1997.

Berlemont, A., Grancher, M.S. and Gouebet, G., Heat and Mass Transfer Coupling Between Vaporizing Droplets and Turbulence Using a Lagrangian Approach, *Int. J. Heat and Mass Transfer*, vol.38, no.17, 1995, pp.3023-3034.

Chen, X.Q., and Pereira, J.C.F., "Numerical Prediction of Nonevaporating and Evaporating Fuels Sprays Under Nonreactive Conditions", *Atomization and Sprays*, Vol.2, 1992, pp.427-443.

Faeth, G. M., "Current Status of Droplet and Liquid Combustion", *Prog. Energy Combust. Sci.*, Vol. 3, 1977, pp. 191-224.

Faeth, G. M., "Evaporation and Combustion of Sprays", *Prog. Energy Combust. Sci.*, vol. 9, 1983, pp. 1-76.

Faeth, G. M., "Mixing, Transport and Combustion in Sprays", *Prog. Energy Combust. Sci.*, vol. 13, 1987, pp. 293-345.

Faeth, G. M., "Spray Combustion Phenomena", Twenty-Sixth Symposium (International) on Combustion/The Combustion Institute, 1996/pp. 1593-1612.

Godsave, G. A. E., "Studies of the Combustion of Drops in a Fuel Spray-the Burning of Single Drops of Fuel, Fourth Symposium (international) on Combustion". Williams & Wilkins, Baltimore, 1953, pp. 818-830.

Gosman, A.D. and Ioannides, E., "Aspects of Computer Simulation of Liquid-Fueled Combustors", AIAA Paper No.81-0323, *AIAA 19th Aerospace Sciences Meeting*, St. Louis, MO, 1981.

Heitor, M. V. and Moreira, A. L. N., "Experiments on Polydisperse Two-Phase Turbulent Jets", ICLASS-94, Rouen, France, Paper XI-5, July 1994.

Hubbard, G.L., Denny, V.E., and Mills, A.F., *Int. J. Heat Mass Transfer*, Vol.16, 1973, pp.1003-1008.

Launder, B.E. and Spalding, D.B., "The Numerical Computation of Turbulent Flows", *Computer Methods in Applied Mechanics and Engineering, vol. 3, 1974, pp. 269-289.*

Law, C. K., "Recent Advances in Droplet Vaporization and Combustion", *Prog. Energy Combust. Sci.*, Vol. 8, 1982, pp. 171-201.

Lefebvre, A. H., "Atomization and Sprays", Hemisphere Pub. Co., New York, 1989.

Leonard, B.P., "A Stable and Accurate Convective Modeling Procedure Based on Quadratic Upstream Interpolation", Computer *Methods in Applied Mechanics and Engineering*, vol. 19, No. 1, 1979, pp. 59-98.

Lockwood, F.C. and Naguib, A.S., "The Prediction of the Fluctuations in the Properties of Free, Round Jet, Turbulent, Diffusion Flames", *Combustion and Flame*, Vol.24, February 1975, pp.109-124.

Ranz, W.E. and Marshall, W.R. Jr., "Evaporation from Drops", *Chem. Eng. Prog.*, Vol.48, 1952, pp. 141-173.

Shirolkar, J. S., Coimbra, C. F. M. and Queiroz McQuay, M., "Fundamentals Aspects of Modeling Turbulent Particle Dispersion in Dilute Flows", *Prog. Energy Combusti. Sci.*, Vol. 22, 1996, pp. 363-399.

Shuen, J.S., Chen, L.D. and Faeth, G.M., "Evaluation of a Stochastic Model of Particle Dispersion in a Turbulent Round Jet", *AIChE Journal*, Vol.19, Jan. 1983, pp.167-170.

Shuen, J.S., Solomon, A.S.P., Zhang, Q.F., and Faeth, G.M., "Structure of a Particle-Laden Jets: Measurements and Predictions", *AIAA Journal*, Vol.23, No.3, pp. 396-404, 1985.

Sirignano, W. A., "Theory of Multi-component Fuel Droplet Vaporization" , *Archives of Thermodynamics and Combustion*, Vol. 9, No.2, 1978, pp. 231-247.

Sommerfeld, M., "Analysis of Isothermal and Evaporating Sprays by Phase-Doppler Anemometry and Numerical Calculations", *Int. J. Heat and Fluid Flow*, Vol.19, 1998, pp.173-186.

Spalding, D. B., "The Combustion of Liquid Fuels, Fourth Symposium (international) on Combustion". Williams & Wilkins, Baltimore, 1953, pp. 847-864.

Saparrow, E.M., and Gregg, J.L., *Trans. ASME*, Vol.80, 1958, pp.879-886.

Tinaut, F.V. (2005). "Performance of Vegetable Derived Fuels in Diesel Engine Vehicles", *Silniki Spanilowe*, No.2/2005 (121).

Watson, K.M., "Prediction of Critical Temperatures and Heats of Vaporization", *Ind. Eng. Chem.*, Vol.23, No.4, 1931, pp.360-364.

Williams, A., "Combustion of Droplets of Liquid Fuels, A Review", *Combustion and Flame*, Vol.21, 1973, pp1-31.

Permissions

The contributors of this book come from diverse backgrounds, making this book a truly international effort. This book will bring forth new frontiers with its revolutionizing research information and detailed analysis of the nascent developments around the world.

We would like to thank Kazimierz Lejda and Paweł Woś, for lending their expertise to make the book truly unique. They have played a crucial role in the development of this book. Without their invaluable contribution this book wouldn't have been possible. They have made vital efforts to compile up to date information on the varied aspects of this subject to make this book a valuable addition to the collection of many professionals and students.

This book was conceptualized with the vision of imparting up-to-date information and advanced data in this field. To ensure the same, a matchless editorial board was set up. Every individual on the board went through rigorous rounds of assessment to prove their worth. After which they invested a large part of their time researching and compiling the most relevant data for our readers. Conferences and sessions were held from time to time between the editorial board and the contributing authors to present the data in the most comprehensible form. The editorial team has worked tirelessly to provide valuable and valid information to help people across the globe.

Every chapter published in this book has been scrutinized by our experts. Their significance has been extensively debated. The topics covered herein carry significant findings which will fuel the growth of the discipline. They may even be implemented as practical applications or may be referred to as a beginning point for another development. Chapters in this book were first published by InTech; hereby published with permission under the Creative Commons Attribution License or equivalent.

The editorial board has been involved in producing this book since its inception. They have spent rigorous hours researching and exploring the diverse topics which have resulted in the successful publishing of this book. They have passed on their knowledge of decades through this book. To expedite this challenging task, the publisher supported the team at every step. A small team of assistant editors was also appointed to further simplify the editing procedure and attain best results for the readers.

Our editorial team has been hand-picked from every corner of the world. Their multi-ethnicity adds dynamic inputs to the discussions which result in innovative outcomes. These outcomes are then further discussed with the researchers and contributors who give their valuable feedback and opinion regarding the same. The feedback is then collaborated with the researches and they are edited in a comprehensive manner to aid the understanding of the subject.

Apart from the editorial board, the designing team has also invested a significant amount of their time in understanding the subject and creating the most relevant covers. They scrutinized every image to scout for the most suitable representation of the subject and create an appropriate cover for the book.

The publishing team has been involved in this book since its early stages. They were actively engaged in every process, be it collecting the data, connecting with the contributors or procuring relevant information. The team has been an ardent support to the editorial, designing and production team. Their endless efforts to recruit the best for this project, has resulted in the accomplishment of this book. They are a veteran in the field of academics and their pool of knowledge is as vast as their experience in printing. Their expertise and guidance has proved useful at every step. Their uncompromising quality standards have made this book an exceptional effort. Their encouragement from time to time has been an inspiration for everyone.

The publisher and the editorial board hope that this book will prove to be a valuable piece of knowledge for researchers, students, practitioners and scholars across the globe.

List of Contributors

Leandro Valim de Freitas
Petróleo Brasileiro SA (PETROBRAS), Brazil
São Paulo State University (UNESP), Brazil

Messias Borges Silva and Carla Cristina Almeida Loures
São Paulo State University (UNESP), Brazil
University of São Paulo (USP), Brazil

Ana Paula Barbosa Rodrigues de Freitas
University of São Paulo (USP), Brazil

Fernando Augusto Silva Marins
São Paulo State University (UNESP), Brazil

M. P. Ashok
Department of Mechanical Engineering, FEAT, Annamalai University, Annamalai Nagar, Tamil Nadu, India

Kazimierz Lejda and Paweł Woś
Rzeszów University of Technology, Poland

Xuelong Miao, Yusheng Ju, Xianyong Wang, Jianhai Hong and Jinbao Zheng
Wuxi Fuel Injection Equipment Research Institute, China

Sibendu Som and Douglas E. Longman
Argonne National Laboratory, USA

Anita I. Ramirez and Suresh Aggarwal
University of Illinois at Chicago, USA

Jorge Barata and André Silva
Aerospace Sciences Department, University Beira Interior, Covilhã, Portugal

Printed in the USA
CPSIA information can be obtained
at www.ICGtesting.com
JSHW011327221024
72173JS00003B/84